RECENT TRENDS IN BIOTECHNOLOGY

EXPLORING THE BIOTECHNOLOGICAL POTENTIAL OF C2 ALTERNATIVE SUBSTRATES IN BIOCATALYSIS AND METABOLIC ENGINEERING

RECENT TRENDS IN BIOTECHNOLOGY

Additional books and e-books in this series can be found on Nova's website under the Series tab.

Recent Trends in Biotechnology

Exploring the Biotechnological Potential of C2 Alternative Substrates in Biocatalysis and Metabolic Engineering

Wenfa Ng

nova science publishers
New York

https://doi.org/10.52305/GTIA8095

Library of Congress Cataloging-in-Publication Data

ISBN: 978-1-68507-555-2

Published by Nova Science Publishers, Inc. † New York

Contents

PREFACE

Metabolic engineering and biocatalysis aim to produce new value-added products from biological routes through careful and deliberate genetic engineering in different microbial chassis with suitable metabolic repertoire. Hitherto, the endeavour has been largely confined to the production of new products from glucose and a few other well characterized substrates. But, as the bioeconomy gains in importance and the world seeks to explore new sources of carbon as feedstocks, possibility exists in developing alternative C2 compounds as feedstocks in contrast to the current reliance on C6 (glucose and fructose), C5 (xylose) and C3 (glycerol) substrates. Two potential C2 alternative substrates explored in this work are ethanol and ethylene glycol. Ethanol is a biofuel whose biotechnological potential has not been fully explored. In the first part of the thesis, the potential of using the NADH generated from a reconstructed ethanol utilization pathway to power a biotransformation reaction is explored in *Escherichia coli*. Experimental results indicate that ethanol utilization delivers higher biotransformation efficiency compared to state-of-the-art glucose dehydrogenase system, which confirmed theoretical predictions of relative efficiency of the two systems. Next, the thesis switches attention to ethylene glycol which could be derived as breakdown products of lignocellulosic waste or from plastic waste. Here, the proposal is to demonstrate the potential of a NADH generating ethylene glycol utilization pathway that terminates at glycolate to support biocatalytic reactions in *E. coli*. Subsequently, the two-step ethylene glycol utilisation pathway would be extended to malate and 2-phosphoglycerate to help another biotechnology workhorse, *Bacillus subtilis*, to grow on ethylene glycol as the sole carbon source. Different connection points to central carbon metabolism would provide a natural experiment to decipher the relative efficiencies of activating tricarboxylic acid cycle and glycolysis in supporting cell growth, and illuminating insights into the relative efficiency and evolutionary importance of different branches of metabolism in *B. subtilis*. Finally, ethylene glycol utilization would be combined with product formation to help *B. subtilis* to synthesis green fluorescent protein using flux from the C2 alternative substrate. This system would provide a platform for examining the efficacy of different ethylene glycol utilisation pathway and entry points in central carbon metabolism in synthesising amino acids and proteins. Overall, this thesis hopes to explore the biotechnological potential of two alternative C2 substrates (ethanol and ethylene glycol) in biocatalysis and metabolic engineering from the substrate utilization perspective.

Keywords: substrate utilization, alternative substrate, cofactor regeneration, alternative protein, metabolic rewiring

ACKNOWLEDGMENTS

I would like to thank my supervisors Dr. Zhou Kang and Prof. Hidajat Kus for their guidance and support. In addition, I would also like to express my gratitude to my labmates for their help and support during my stay in the lab. Sincere thanks also go out to my lab officer, Mr. Lim You Kang, for making the logistic workflow of labwork manageable and providing help and guidance whenever something goes wrong in the lab. I would also like to thank Liang Hong and Lu Hongyuan for providing critical plasmids with ethanol utilisation pathway genes and cpsADH, respectively, to start off my research. Finally, I thank my family for their support and encouragement.

Chapter 1

INTRODUCTION

1. METABOLIC ENGINEERING FROM THE SUBSTRATE UTILIZATION PERSPECTIVE

Metabolic engineering involves the carefully choreographed series of genetic manipulations that endow microbial cells with new-found capabilities at enhanced conversion of substrates into products. The focus till now has predominantly been on developing different metabolic rewiring strategies for the production of different metabolites. Hence, different studies abound in understanding how different heterologous metabolic pathways could be used in sync with parts or branches of endogenous metabolism in enabling the production of desired products. With the effort of the global metabolic engineering community, the product space in metabolic engineering has significantly expanded across different product classes to even those previously thought to be intractable from a biological perspective. Thus far, glucose is the enabling substrate for the production of a multitude of products given its ability at activating expression of multiple branches of endogenous metabolism through the glucose dependent regulon. This is known as the product perspective in metabolic engineering.

In contrast, relatively little is known about the use of alternative substrates in enabling bioproduction [1, 2]. Such alternative substrates do not refer to sugars-based substrates such as fructose, xylose, arabinose or glycerol. Substantial body of literature exists in describing their use in enabling production of useful metabolites and intermediates [2-4]. What constitutes the focus of this proposed thesis is a class of substrates not naturally assimilated by common biotechnological workhorses such as *Escherichia coli, Bacillus subtilis,* and *Saccharomyces cerevisiae*. More importantly, the substrates of interest should not have natural regulon in the microbial chassis chosen. Regulon, in this case, refers to the ensemble of genes sensitive to the presence of or whose expression are regulated by the substrate of interest. Hence, the metabolic engineering perspective espoused in this work seeks to create a platform for the utilization of alternative substrates, that when suitably connected to central carbon metabolism, could endow cell growth on the substrate as sole carbon source as well as enabling the channelling of substrate flux for the production of useful products.

The alternative substrates of interest in this proposed thesis research are ethanol and ethylene glycol. Both are unconventional carbon sources for microorganisms, particularly in the case of ethylene glycol. Ethanol, on the other hand, is a biofuel whose biotechnological potential has not been fully explored. Specifically, its use as a substrate for growth and powering the endogenous metabolism of cells remain incompletely characterized and understood given that only a few organisms have been shown to achieve growth on the substrate [5, 6]. Although ethanol is known to be catabolized in two important biotechnological workhorses, *E. coli* and *B. subtilis*, its use as a substrate remains limited as it does not enjoy multiple connections to endogenous central carbon metabolism. The latter point is important as availability of enzymes that could catabolize ethanol to various intermediates in central carbon metabolism would allow ethanol to be connected to different branches of endogenous metabolism that, in turn, assures pathway resilience.

In *E. coli*, ethanol could be catabolized into acetyl-CoA through the intermediate molecule, acetaldehyde [7]. Given the importance of acetyl-

CoA in growth promotion, ethanol metabolism could be connected to central carbon metabolism and cell growth. But, the focus of this thesis is not on utilization of ethanol for promoting cell growth. Rather, the first subtopic of this thesis looks into the use of NADH generated by the two sequential reactions that push ethanol flux to acetyl-CoA for powering a biotransformation reaction mediated by an alcohol dehydrogenase. The two reactions are mediated by heterologous enzymes that converts ethanol to acetyl-CoA through the same reaction sequence as the endogenous adhE enzyme in *E. coli*. Specifically, during the conversion of ethanol to acetyl-CoA, NADH is generated from the NAD^+ dependent oxidoreductase mediating the two steps conversion. The NADH generated could subsequently be used to power a NADH dependent conversion of a ketone (butanone) into alcohol (butanol). Value of this proposed NADH cofactor regeneration scheme arises from the higher NADH regeneration efficiency of ethanol compared to glucose, the latter via the glucose dehydrogenase system. On a per ethanol molecule level, each ethanol molecule could regenerate two NADH molecule compared to one NADH molecule by the glucose/glucose dehydrogenase system. Hence, at the theoretical level, ethanol utilization pathway should be more efficient than the glucose/glucose dehydrogenase system in regenerating NADH. The only factor disfavouring the use of ethanol is its higher price compared to glucose.

With respect to biology, ethylene glycol is an unnatural two carbon substrate that could not be utilized by many organisms. But, the substrate may rise in value and importance given the availability of two sources of input: (i) partial degradation of lignocellulosic waste, and (ii) enzymatic degradation of polyethylene terephthalate (PET) plastic waste. Hence, ethylene glycol may be a good source of carbon for building biomass or production of useful products. Naturally, utilization of ethylene glycol requires building a pathway that connects the substrate to central carbon metabolism. In *E. coli*, one possibility lies in the sequential two-step conversion of ethylene glycol to glycoaldehyde and through to glycolate [8]. This heterologous pathway does not connect ethylene glycol to one of the key intermediates in central carbon metabolism. But, the pathway remains useful given its ability at generating two NADH molecules per turn of the cycle, which can be utilized to support a biotransformation reaction mediated by an alcohol dehydrogenase. The described system is similar to that for ethanol utilization where the biotransformation substrate is butanone with butanol as product. Value proposition of this project comes from exploring another facet in which ethylene glycol utilization could benefit biocatalysis and metabolic engineering, especially where NADH produced in earlier part of the ethylene glycol utilization process could be used in latter part of a heterologous pathway producing a useful product.

To better explore the biotechnological potential of ethylene glycol, a metabolic pathway guiding its utilization must connect to a more important intermediate in central carbon metabolism such as those in tricarboxylic acid (TCA) cycle or glycolysis. A more important intermediate is defined by either the number of connections such a metabolite has with other pathways as well as the efficiency in which metabolic flux could be drained from the connecting intermediate, thereby, allowing metabolic flux and carbon to flow seamlessly between substrate and product. This segment of the thesis work aims to use the biotechnological workhorse and Generally regarded as safe (GRAS) microbe, *Bacillus subtilis*, as microbial chassis to build an ethylene glycol utilization pathway that converts ethylene glycol to biomass. To this end, ethylene glycol would be guided beyond glycolate to either reach malate (TCA cycle) and 2-phosphoglycerate (glycolysis) to compare the relative efficiency in which different metabolic connection points could help promote biomass formation. Knowledge that could emanate from this work includes the demonstration that ethylene glycol could serve as a sole carbon source for a biotechnological workhorse, which expands the available substrate scope for biotechnology. Secondly, the work also offers insights into how different ways of connecting a heterologous substrate utilization pathway to central carbon metabolism could impact on biomass formation efficiency.

The final segment of this thesis work aims to further explore the possibility of using ethylene glycol as substrate to power the production of a useful molecule, in this case, a recombinant protein. To this end, pathways emanating from ethylene glycol must connect to metabolites in central carbon metabolism to channel flux for producing all twenty amino acids. Key intermediates in cellular metabolism that feed amino acid synthesis are oxaloacetate, αketoglutarate, pyruvate, 3-phosphoglycerate and erythrose-4-phosphate, which are distributed in tricarboxylic acid cycle, glycolysis and pentose phosphate pathway. The key hypothesis that would be addressed in this segment of work is to elucidate if different entry points to central carbon metabolism affect the proportion of amino acids produced and the relative efficacy of recombinant protein production.

To synthesize a recombinant protein using ethylene glycol as feedstock would demand that ethylene glycol be connected to a metabolite in central carbon metabolism that could activate all necessary pathways for the synthesis of amino acids. This requires the connecting metabolite to be well-connected to other pathways. One candidate is acetyl-CoA. Another two candidate metabolites are malate and 2-phosphoglycerate. The ultimate choice amongst the three would come from experimental studies examining the relative efficacy of the 3 pathways emanating from ethylene glycol in enabling recombinant protein production. Green fluorescent protein would be the model protein and the production host would be *B. subtilis*. Choice of *B. subtilis* comes about due to its ability to secrete many proteins and metabolites into the extracellular space [9, 10]; thereby, facilitating protein production and downstream processing. Knowledge gained from this project should point to any technological gap in developing ethylene glycol utilisation for protein production in *B. subtilis*. More importantly, the results obtained should also inform on the choice of entry point in central carbon metabolism for guiding ethylene glycol flux to help improve heterologous protein production.

Overall, this thesis proposal aims to illuminate strategies that have been or will be taken to explore the biotechnological potential of two C2 alternative substrates in biocatalysis and metabolic engineering. Choice of the alternative substrate does not come from convenience or an abundance of extant feedstock. Rather, ethanol and ethylene glycol both possess deep biotechnological and scientific appeal given their C2 nature of a two-carbon compound. While life can be construed to be built upon C6 compounds such as the six-carbon glucose and fructose with C5 compound (xylose) and C3 compound (glycerol) in supporting roles, the importance of C2 compounds in the evolution of metabolic pathway is relatively underappreciated. It is hoped that through exploring the biotechnological potential of C2 alternative substrates that we can arrive at some initial understanding of how life evolve to neglect C2 compounds in preference for C6 and C5 compounds. In developing strategies to plug C2 compounds such as ethanol and ethylene glycol conveniently into central carbon metabolism, the enduring desire should be to traverse a path that may have been forgotten by evolution, or which has been deemed to be extraneous and unimportant.

2. Rationale for Choosing C2 Metabolites as Substrates

Why take the unusual route to work on C2 compounds for biocatalysis and metabolic engineering when C6 compounds such as glucose has been known to activate many metabolic pathways through inducing the expression of relevant enzymes? The answer is three-fold. Firstly, glucose being the preferred substrate for many microorganisms and biotechnological workhorse has received much attention in biocatalysis and metabolic engineering. Indeed, the C6 compound glucose is the standard by which new heterologous metabolic pathways are planned, construed and developed. This then leaves open a significant gap in which metabolic flux flow is initiated in any metabolic engineering or biocatalysis project. Specifically, flux typically flows top-down from glucose to acetyl-CoA in glycolysis. Relatively little is known about the efficiency in which flux could flow from a C2 substrate such as ethanol and ethylene glycol into central carbon metabolism. Although modern omics tool such as RNA-seq could provide glimpses of the answers through comparison of relative abundance of different metabolic enzymes in different branches of central carbon metabolism, much remains unknown and awaits experimental verification.

Secondly, C2 compounds offer a shorter route to a key growth promoting metabolite, acetyl-CoA, in central carbon metabolism. Specifically, it takes two sequential enzymatic conversion to transform ethanol to acetyl-CoA, which helps connect ethanol to tricarboxylic acid (TCA) cycle and lipid synthesis. Experimental assays interrogating growth performance of *E. coli* on M9 medium supplemented with ethanol confirmed that *E. coli* could grow on ethanol with a heterologous two-step pathway linking ethanol to acetyl-CoA. Ethylene glycol, on the other hand, presents a more difficult challenge. One possibility is to convert ethylene glycol into glycoaldehyde and subsequently to glycolate [8]. However, this route does not connect ethylene glycol to any major intermediate in central carbon metabolism, which opens up a vast space for creative metabolic engineering in linking ethylene glycol to central carbon metabolism, in particular, acetyl-CoA. Doing so, a large space of metabolic reaction possibilities could be probed for their feasibility in linking ethylene glycol to acetyl-CoA and, by extension, central carbon metabolism, which in addition to augmenting our metabolic reaction toolkit for engineering, also help illuminate hitherto unexamined reaction possibilities *in vivo*.

Thirdly, C2 substrates represent a class of carbon molecules that could serve as building blocks for more complex molecules. Specifically, in the process of building biomass from C2 substrates, anabolic reactions would definitely dominate over catabolic reactions in the overall reaction scheme. Thus, C2 substrates as biomass building blocks would not incur significant amount of carbon loss, thereby, making the overall metabolic pathway more atom efficient. On the other hand, traditional metabolic engineering based on glucose as substrate would incur substantial amount of carbon loss as carbon dioxide due to the need to catabolize C6 glucose to C3 and C2 molecules, where some decarboxylation reactions release the extra carbon as carbon dioxide, which represents a loss to the cell and decrease the overall product yield.

Chapter 2

DEVELOPING ETHANOL UTILISATION FOR NADH REGENERATION IN BIOCATALYSIS

1. INTRODUCTION

Ethanol is more commonly thought of as a product than a substrate. But, some microorganisms such as *Escherichia coli* and *Bacillus subtilis* do possess the capability at catabolizing ethanol as a substrate (from experimental data collected in this work). This comes about likely due to the co-evolution of both species as gut commensals in human where periodic infusions of ethanol in wine and alcoholic beverages challenged the gut microbiota at developing solutions to tackle an existential challenge given that high ethanol concentration could kill bacterial cells.

The solution that evolved in *E. coli* takes the form of a two-step pathway that first converts ethanol into an acetaldehyde, and subsequently to acetyl-CoA [7]. From first inspection, the pathway is efficient in connecting ethanol to a key metabolite in central carbon metabolism that feeds into both tricarboxylic acid (TCA) cycle and lipid synthesis, thereby, offering opportunities for the use of ethanol as a sole carbon source for supporting growth. Experimental growth assay investigating the issue confirms that ethanol could serve as the sole carbon source for promoting the growth of *E. coli* BL21 (DE3) in M9 minimal medium. The enzyme responsible for the above-mentioned ethanol catabolism pathway is adhE, which is a multidomain molecular machine capable of catalyzing both steps of the ethanol catabolism pathway [7].

This work sought to emulate the endogenous adhE enzyme by reconstructing the ethanol utilisation pathway as a plasmid-based heterologous system suitable for use in different biotechnological workhorses such as *E. coli, B. subtilis* and *S. cerevisiae*. To this end, the pathway was envisioned to be of two sequential steps each catalyzed by its respective enzyme. In particular, the first step is catalyzed by an alcohol dehydrogenase (adh2) from *S. cerevisiae* and the second step catalyzed by an acetaldehyde dehydrogenase (ada) from *Dickeya zeae*. Both steps of the pathway use NAD^+ dependent enzymes and thus regenerates NADH, which could be tapped for either biocatalysis or metabolic engineering applications.

NADH that emanates from the two-step heterologous ethanol utilization pathway (EUP) would be tapped for powering a biotransformation reaction converting a ketone to an alcohol mediated by an alcohol dehydrogenase, cpsADH from *Candida parapsilosis*. Current state-of-the-art in NADH regeneration uses glucose dehydrogenase (GDH) that converts glucose to gluconolactone to perform NADH regeneration [11]. From the theoretical perspective, the EUP pathway offers advantages of a higher efficiency in NADH regeneration compared to glucose/glucose dehydrogenase system given its ability to regenerate two NADH per ethanol molecule. In contrast, each glucose molecule could only regenerate one NADH through the GDH system. Although ethanol is more expensive than glucose as co-substrate in any biotransformation system, the EUP system remains valuable for investigation with regards to its NADH regeneration efficiency given that its terminal product, acetyl-CoA could feed into biomass formation processes and thus creates a driving force in growing cell biotransformation. Capability of generating a driving force would further improve EUP's NADH regeneration efficiency and strengthen its case for use in powering biotransformation reactions in biocatalysis.

Cofactor regeneration systems could operate both *in vivo* and *in vitro*. Given the high cost of purified enzymes, the NADH regeneration system designed in this study is catered for *in vivo* whole-cell biocatalysis. The microbial chassis chosen is *E. coli* BL21 (DE3) that harbours two plasmids: (i) one carrying cpsADH, and (ii) another carrying EUP pathway genes. Both resting and growing cell biotransformations were carried out to assess the efficacy of NADH regeneration by EUP pathway. Here, resting cells biotransformation refers to biotransformation carried out in the absence of a growth medium but with provision of cosubstrate ethanol and biotransformation substrate. On the other hand, growing cell biotransformation refers to cells grown in a growth medium carrying out biotransformation with the aid of the co-substrate and main substrate.

Experimental results revealed that *E. coli* harbouring the reconstructed EUP pathway on a plasmid could grow on ethanol, which meant that the two-step EUP pathway is functional. More importantly, in biotransformation assays converting butanone to 2-butanol, product yield was enhanced with use of EUP for NADH regeneration relative to glucose/GDH system. This validated theoretical predictions that the EUP pathway is more efficient in regenerating NADH compared to glucose/GDH system. Furthermore, growing cells were shown to enable higher biotransformation efficiency compared to resting cells presumably due to the driving force generated by cell growth that push ethanol flux to acetyl-CoA. Finally, gene expression pattern resultant from cultivation in different growth media either prior to or during biotransformation also impact on biotransformation efficiency. In this case, M9 medium supplemented with 10 g/L ethanol was shown to enable higher biotransformation efficiency than LB medium in both resting cell and growing cell biotransformation.

The overall picture that emanates from the experimental results collected point to ethanol utilization as a viable platform for regenerating NADH in whole cell biocatalysis converting butanone to 2-butanol in *E. coli*. Given that it is a plasmid-based platform, the twogene EUP pathway could be readily transplanted to other species for enabling NADH regeneration in support of a NADH requiring biotransformation reaction after adjustment in plasmid replication origin and promoters to tailor to the genomic context of individual species. Although EUP is more efficient than the state-of-the-art glucose dehydrogenase system in regenerating NADH, higher cost of ethanol may still tilt the balance in favour of glucose dehydrogenase system. However, the scientific questions explored in this project should provide a firm foundation for future endeavours in tapping the NADH regeneration potential of EUP pathway such as in metabolic engineering applications where ethanol is the substrate providing the carbon source for building a target product molecule. In this case, downstream reactions whose enzymes require NADH would be able to use the NADH generated by EUP to achieve redox balance in the cell; thereby, ensuring a conducive cellular environment for attaining higher product yield and titer. Hence, ethanol/EUP system would then become a tool in the toolbox of metabolic engineers designing new pathways linking substrate to desire product where EUP provide a source of acetyl-CoA and NADH.

2. Theoretical Underpinnings of the Project

Ethanol utilization pathway lies at the heart of this whole-cell NADH regeneration project, and would be described in this segment. Being a C2 molecule, ethanol could interface with central carbon metabolism using a C2 scaffold at acetyl-CoA. Two sequential enzymatic steps converting ethanol first to acetaldehyde and subsequently to acetyl-CoA would connect ethanol expeditiously to central carbon metabolism (Figure 1). Although such metabolic connection is useful for metabolic engineering, it may also be useful for biocatalytic application, especially if the pathway regenerates a redox cofactor such as in this case.

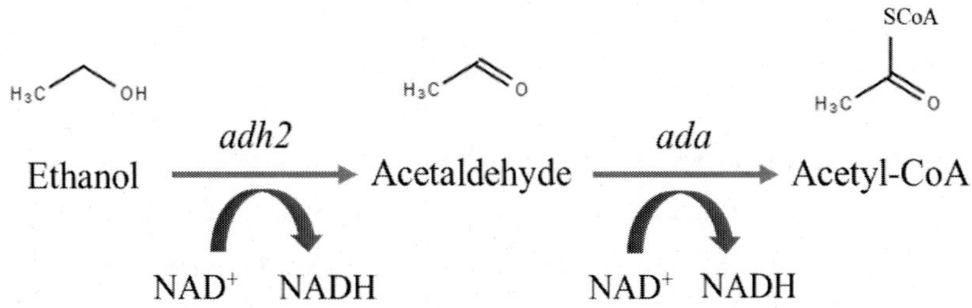

Figure 1. Architecture of ethanol utilization pathway comprising two genes and enzymatic steps mediating the stepwise conversion of ethanol to acetaldehyde and on to acetyl-CoA.

In the first step of EUP, adh2 from *S. cerevisiae* converts ethanol to acetaldehyde with the generation of one NADH from NAD^+. Subsequently, the toxic intermediate acetaldehyde must be converted efficiently to acetyl-CoA through ada from *Dickeya zeae*. Similar to the first step of EUP, the second enzymatic reaction is also cofactor dependent and converts NAD^+ to NADH. Thus, in aggregate, two NADH molecules are generated per turn of the cycle or per ethanol molecule. How does this compare with the state-of-the-art glucose dehydrogenase system?

The glucose dehydrogenase system comprises a single enzymatic step converting glucose to gluconolactone (Figure 2). In the process, a NAD^+ molecule is converted into NADH, which endows the reaction cofactor regeneration properties. Compared to EUP, the glucose dehydrogenase system suffers from a lower theoretical regeneration of NADH per molecule or per turn of the cycle. Hence, EUP should enjoy advantages of higher NADH regeneration efficiency compared to glucose dehydrogenase even though ethanol is more costly as a substrate relative to glucose.

The concept of cofactor regeneration comes about in the early 2000 when researchers in the biocatalysis field encounter problems with regards to provision of costly cofactors such as NADH or NADPH for oxidoreductase enzymes in biotransformation reactions. Idea for regenerating the costly redox cofactor was born through integrating a cofactor regeneration reaction with the target biotransformation reaction (Figure 3). Briefly, in cofactor regeneration schemes, a cofactor requiring reaction converting substrate 1 to product 1 is coupled to a cofactor regenerating reaction converting a second substrate to a second product. Such a cofactor regeneration scheme avoids the provision of costly cofactor, where the cost of regenerating the cofactor is transferred to that of the second substrate. However, provision of a second substrate and generation of a new product may complicate the downstream purification of the target product, thereby, adding to overall process cost.

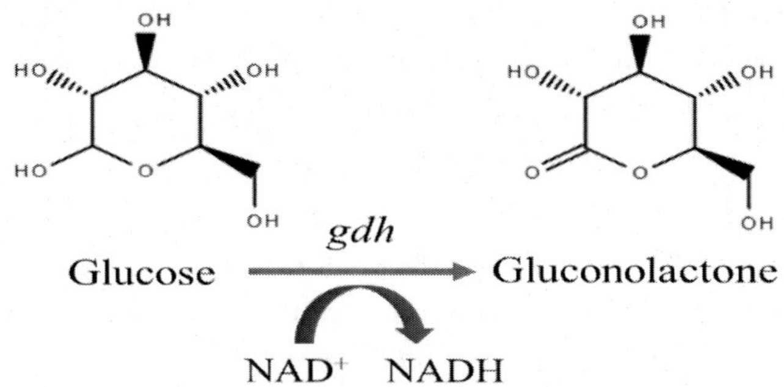

Figure 2. Glucose dehydrogenase NADH regeneration system is buttressed by a single enzymatic step converting glucose to gluconolactone that helps regenerate a single NADH molecule.

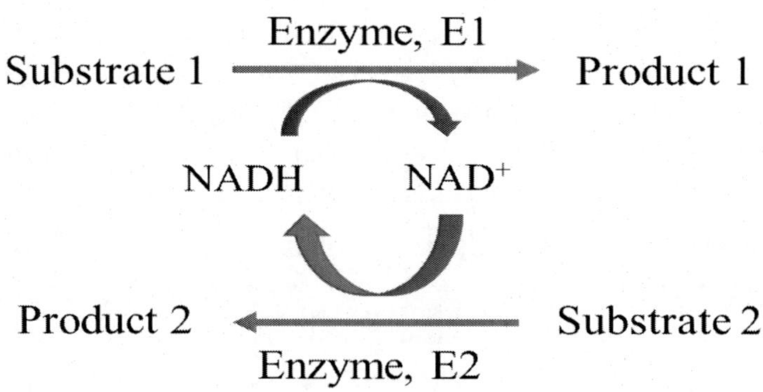

Figure 3. Conceptual basis of cofactor regeneration schemes in biocatalysis. The approach works by coupling a cofactor requiring reaction with another one that regenerates the cofactor. In this way, there is no need to supply expensive cofactors. However, addition of another substrate and generation of a product may complicate downstream purification of the target product.

The project focuses on demonstrating the utility of the EUP pathway in enabling NADH regeneration, diagnosing any bottlenecks that may hamper optimal cofactor regeneration efficiency, as well as evaluating its potential against the state-of-the-art NADH regeneration system: glucose dehydrogenase. To this end, a biotransformation reaction converting a ketone to an alcohol is selected to help evaluate the performance of each NADH regeneration system, where more NADH regeneration would enable higher yield to be achieved for a NADH dependent reaction. An alcohol dehydrogenase from *C. parapsilosis* (cpsADH) was selected for catalyzing this biotransformation reaction given its broad substrate specificity that afford it great versatility in biocatalytic applications such as producing different products from different substrates.

To show that the EUP pathway is generic and is capable of enabling higher product yield for different biotransformation substrate, two substrates (acetophenone and butanone) were chosen for this work. The two substrates differ critically in their hydrophobicity and thus necessitate the use of different biotransformation system. Specifically, acetophenone is highly hydrophobic and requires a secondary organic phase for effective dissolution. On the other hand, butanone is highly hydrophilic and could be dissolved in the same aqueous phase as the cells, and thus, does not need a secondary organic phase. One advantage of working with the two dichotomously different substrates is the ability of testing the NADH regeneration efficiency of the EUP pathway under different process conditions (i.e., single phase or two phase biotransformation systems), which holds relevance to final industrial implementation. For example, a two-phase biotransformation system could only be efficient in substrate conversion provided a driving force exists in pulling acetophenone to the aqueous phase. Such a driving force could be consumption of NADH by cpsADH.

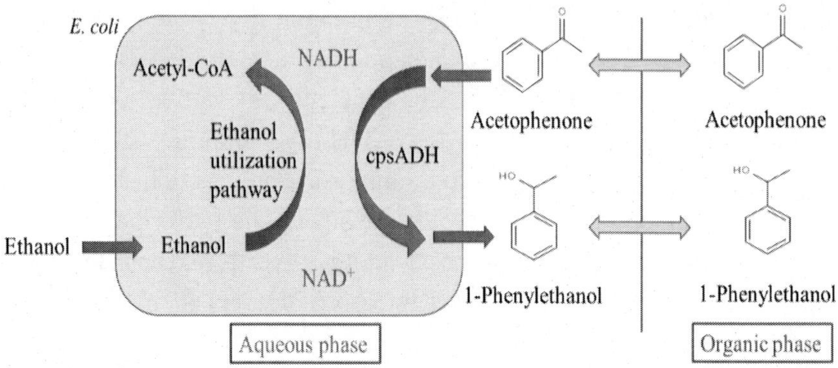

Figure 4. Diagram illustrating the partition of substrate (acetophenone) and product (1-Phenylethanol) between the organic phase (hexadecane) and aqueous phase where the cells reside in two-phase biotransformation.

Overall, a typical two-phase biotransformation system is as depicted in Figure 4. The organic phase, hexadecane, serves as a reservoir for the storage of the hydrophobic substrate, acetophenone, which diffuses to the aqueous phase in small quantities given its poor solubility in water. Similarly, the product, 1-Phenylethanol, would diffuse from the aqueous phase to the organic phase. In whole-cell biocatalysis, acetophenone would diffuse into the cell and be converted to form 1-Phenylethanol, which diffuses out of the cell. NADH needed by cpsADH for the biotransformation reaction is supplied by the ethanol utilization pathway that converts ethanol to acetyl-CoA. Similar to acetophenone, the co-substrate, ethanol, is imported by the cell (check this point). In terms of metabolite analysis, acetophenone and 1-Phenylethanol were detected by gas chromatography-mass spectrometry (GC-MS), while ethanol and acetate were detected by high performance liquid chromatography (HPLC).

As mentioned, biotransformation reactions could be carried out either in single phase or two-phase system. Different from a two-phase biotransformation system, a single phase biotransformation could dissolve all the substrates, co-substrates and products, which suffice for systems utilizing the highly water soluble substrate and product: butanone and 2-butanol, respectively (Figure 5). Similar to the case for acetophenone and

1-Phenylethanol, butanone would be taken up by the cell during whole-cell biocatalysis. After conversion to 2-butanol by cpsADH enzyme, the product would diffuse out of the cell (check this out). In a similar fashion, the co-substrate, ethanol, would be imported by the cell and be converted to acetyl-CoA via the ethanol utilization pathway, and in the process generate NADH.

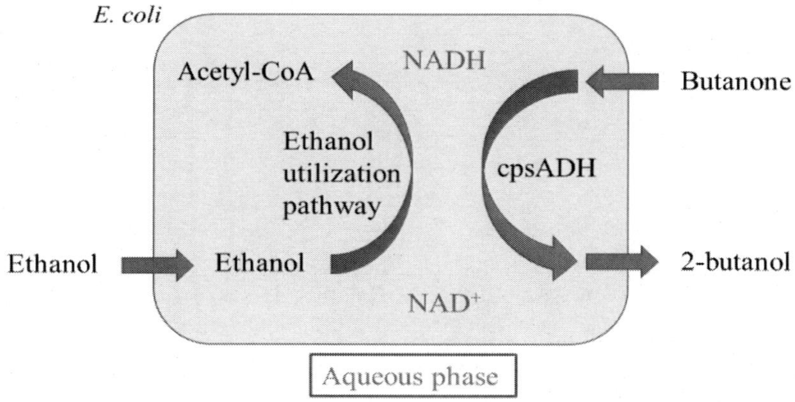

Figure 5. Diagram illustrating the principles underlying one a single-phase biotransformation system work. In general, water is the solvent and all the substrates and products should be highly water soluble.

A key hypothesis examined in this project is whether higher theoretical efficiency in regenerating NADH would afford EUP pathway higher product yield compared to that achievable via the glucose/GDH system. Overall, the NADH regeneration scheme with butanone as substrate is as follows in Figure 6. EUP has a higher theoretical yield for NADH compared to GDH, and should provide for higher product yield. But the comparison is restricted to the terminal product of the heterologous pathway. Given that both acetyl-CoA and gluconolactone could be plugged into endogenous metabolism where there may be NADH requiring steps, the overall efficiency in NADH regeneration in a whole-cell biocatalysis system is likely to be lower with corresponding lower than expected product yield for EUP pathway. However, the overall trend in which EUP should enable higher product yield and biotransformation efficiency over the GDH system should be correct and observable in experiments.

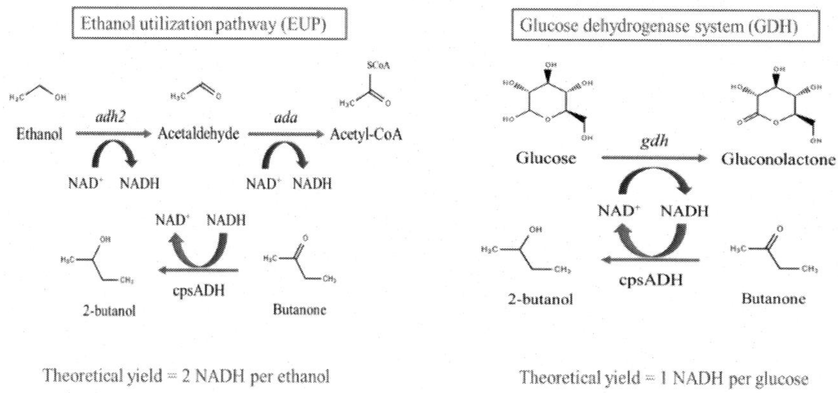

Figure 6. Comparison between EUP and GDH in NADH regeneration for powering butanone biotransformation to 2-butanol. EUP has a higher theoretical yield for NADH than GDH and should provide higher yield of product, 2-butanol in a single-phase biotransformation system.

In this project, both resting cell and growing cell biotransformation were utilized to examine the efficiency in which EUP pathway could regenerate NADH. Regeneration efficiency is evaluated by analysing the concentration of the product, 2-butanol, given the inherent difficulty in accurate measurement of NADH concentration in a whole-cell biocatalytic system. Resting cells as a system bears resemblance to industrial biocatalytic system where cells are denied growth nutrients after the initial growth phase. In the production phase, only the main biotransformation substrate and a co-substrate for biotransformation is used. Cells are thus in the "resting" state and could not grow. Relatively little is known about the physiology of resting cells in the open literature (cite). One important assumption underlying the use of resting cells is that such cells would not waste carbon and energy on further growth process (Figure 7). Rather, energy and carbon could be dedicated to biocatalysis and product formation. Hence, the resting cell concept is an important pillar in modern industrial biocatalysis operation.

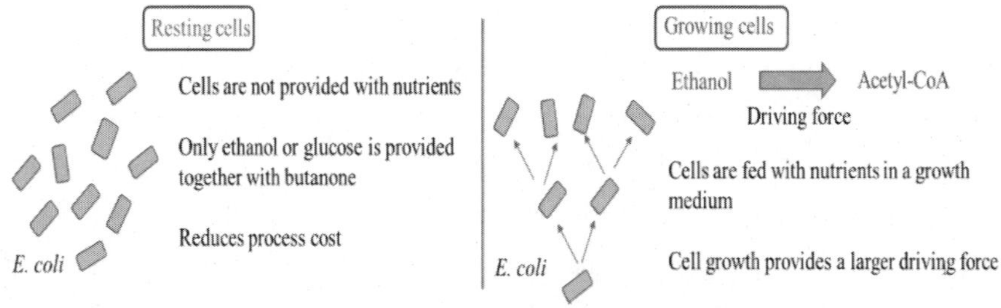

Figure 7. Comparison between the key characteristics of resting cell and growing cell biotransformation. In particular, presence of a growth-promoting metabolite in the heterologous pathway affords the use of cell growth to drive metabolic flux through the pathway via consuming the growth-promoting metabolite, which in the case of the EUP pathway is acetyl-CoA.

But, in some circumstances, growing cells could provide additional driving force enabling higher substrate consumption and NADH regeneration. The concept underlying growing cells is most commonly applied in metabolic engineering, where glucose serves as both carbon and energy source to growth processes to generate the building block in aid of product and biomass formation. To tap on the concept of growing cell in biotransformation, a growth promoting metabolite must be present in the heterologous pathway. Such a metabolite, acetyl-CoA, exists in the heterologous EUP pathway of this project (Figure 7). With depletion of acetyl-CoA pool in the cell due to its use in growth processes, there would be a forward shift in the equilibrium position of the two reversible reactions in the pathway that guide more ethanol flux to be converted to acetyl-CoA. With greater consumption of ethanol comes higher regeneration of NADH and corresponding greater product yield. Given that the EUP pathway possess a pathway intermediate that also serves as a key growth promoting metabolite for the cell, it would be interesting to examine if growing cell would trump resting cell in enabling higher biotransformation efficiency. From the theoretical perspective, growing cell should perform better.

Growth medium plays a critical role in providing sufficient nutrients to facilitate high cell density cultivation as well as activating expression of genes suitable for allowing high efficiency biotransformation. The latter effect is the primary motivation for the use of two different growth medium in this project: LB Miller medium and M9 ethanol medium.

Specifically, it is a scientific curiosity with respect to this project to investigate how cell growth in different growth media might affect gene expression, and whether the gene expression pattern obtained would be conducive for enabling high biotransformation efficiency. Theoretically, gene expression pattern play an important role in biotransformation experiments given that many biocatalytic and metabolic engineering pathways taps and connects to endogenous metabolism. But, the effect is likely to be more prominent in growing cell biotransformation. Specifically, cell growth requires the activation of a larger set of genes than the heterologous pathway of interest in biocatalysis and metabolic engineering, which meant that greater opportunities exist for different growth medium and gene expression pattern to influence final biotransformation obtained. The crux of the issue is that different growth medium activates different set of genes and cellular processes that may ultimately affect how the intermediates and products of a heterologous pathway are consumed, which, in turn, affect biotransformation efficiency. In the case of M9 ethanol medium, the goal is to adapt the cells to grow on ethanol in order to help facilitate greater ethanol consumption and higher biotransformation efficiency. This effect should be present in both resting cells and growing cells given that ethanol consumption in the cultivation phase should induce some changes in *E. coli* gene expression pattern which could be translated to different product concentration and biotransformation efficiency.

The goal of this project is to build and test a plasmid-based NADH regeneration system driven by ethanol. Hence, the two genes that constitute the EUP pathway were encoded by a plasmid with ampicillin as antibiotic resistance gene marker and pMB1 replication origin. An autoinducible promoter from *E. coli*, pthrC3, controls the expression of the two genes on the plasmid (Figure 8a). On the other hand, the cpsADH enzyme is carried by another plasmid with spectinomycin as antibiotic resistance gene marker and p15A replication origin (Figure 8b). Similar to EUP pathway, expression of cpsADH is also under the control of pthrC3 promoter. Since both plasmids have different origins of replication and antibiotic resistance markers, they could be stably maintained in a transformed *E. coli* cell.

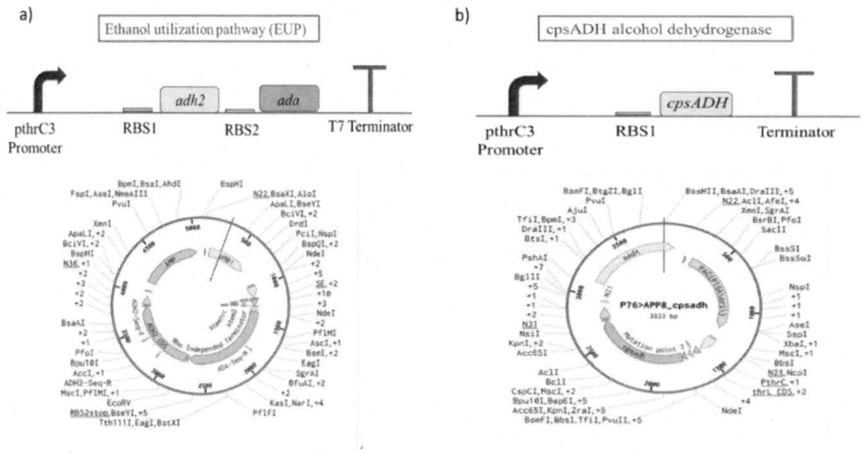

Figure 8. Gene expression cassette depicting the promoters, genes and terminators used in the expression of ethanol utilization pathway genes and cpsADH gene in two different plasmids that are co-transformed into *E. coli*. a) Ethanol utilization pathway plasmid, b) plasmid carrying cpsADH.

Experiments with resting cells and growing cells follow the same general protocol where cells were first cultivated in either LB Miller or M9 ethanol medium. After 18 hours of cultivation, cells were harvested and either suspended in phosphate buffer (resting cells) or growth medium (growing cells) for biotransformation. Anaerobic conditions were used for biotransformation to simulate bioprocessing conditions in the industry where oxygen is not sparged into the bioreactor. Biotransformations were conducted in small volumes of 0.5 mL in HPLC vials at 37°C for 44 hours. Metric for assessing biotransformation efficiency and effectiveness of NADH regeneration is yield per unit OD. The general workflow for biotransformation is as shown in Figure 9.

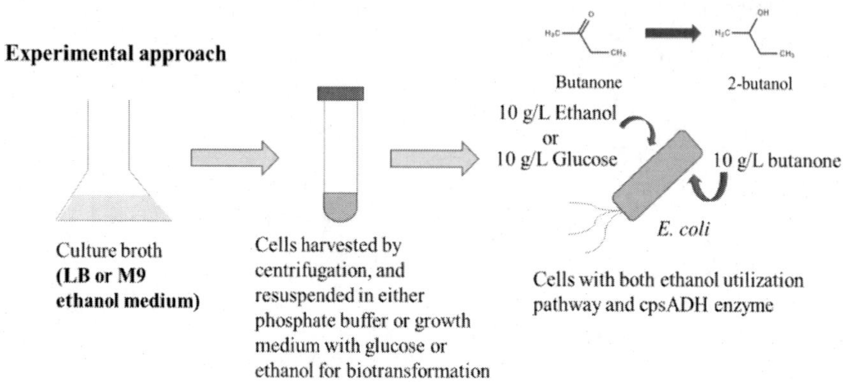

$$Yield\ per\ unit\ OD = \frac{butanol\ conc}{Initial\ butanone\ conc} \times \frac{100}{Final\ optical\ density}$$

Figure 9. Schematic diagram illustrating the general experimental workflow for characterizing the performance of different EUP pathway variants or experimental conditions.

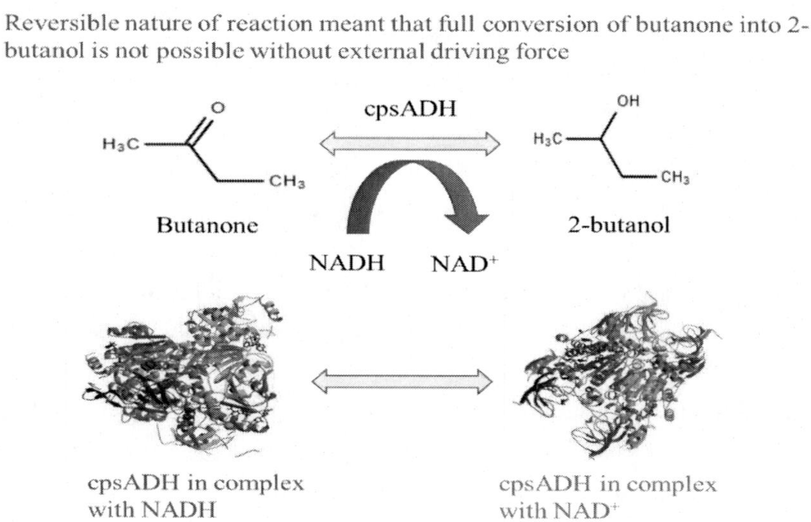

Figure 10. Reversible conversion of butanone to 2-butanol could not run to completion without external driving force. Substantial conformation change in cpsADH enzyme in complex with NAD^+ and NADH highlights that large structural changes are needed during the biocatalytic step.

With regards to the enzymes selected for this experimental system, the main biotransformation enzyme, cpsADH, is a carbonyl reductase from *Candida parapsilosis*. It is a NAD^+/NADH dependent oxidoreductase capable of interconverting a ketone to an alcohol and vice versa. Thus, it is a reversible enzyme, and it is expected that difficulty would exist in obtaining high conversion of substrate to product without external driving force such as an influx of NADH from the EUP pathway such as that anticipated during growing cell biotransformation. From experimental tests, cpsADH has a broad substrate scope and is thus expected to be a versatile enzyme suitable for a variety of applications. Furthermore, experimental test conducted in this study has confirmed that cpsADH has a strict requirement for NADH to facilitate the conversion of butanone to 2-butanol. Shown in Figure 10 is the crystal structure of cpsADH in complex with NADH or NAD^+.

Overall, it is hoped that the preceding sections on different aspects of the project would provide the reader with an understanding of the theoretical feasibility of the project, how the different hypothesis are formulated and the experimental methods chosen to test the ideas.

3. Literature Review

Many industrial biotransformations employ oxidoreductase for converting an intermediate into a useful product. This is in the realm of biocatalysis. Typically, these oxidoreductases are cofactor dependent and require a constant supply of cofactor for function, which are typically NADH and NADPH reducing equivalents [12]. Providing stoichiometric amounts of these expensive cofactors would be too costly for a bioprocess. Hence, cofactor regeneration reactions are set up to support biotransformation reactions requiring cofactors through *in situ* regeneration of cofactors. The process works by turning the oxidized cofactors into its reduced form via enzymatic reaction. In this way, the needed cofactor is regenerated and there is no need to provide it stoichiometrically. Another benefit of the cofactor regeneration scheme is the shifting of cost of providing cofactors to that of the co-substrate that powers cofactor regeneration. This then place the onus on the discovery of new cofactor regeneration schemes on the utilisation of cheap substrate capable of highly efficient regeneration of needed cofactors.

This project focuses on NADH cofactor regeneration for whole-cell biocatalysis where both the biotransformation and cofactor regeneration reactions occur in the cell. In this way, the cofactor regenerated remains in the cell to support the biotransformation reactions. Currently, glucose dehydrogenase is the most popular and state-of-the-art NADH cofactor regeneration system, even though other systems that utilize, for example, glycerol dehydrogenase [13, 14], formate dehydrogenase [15], and NADH oxidase [16, 17] could also do the job. Glucose dehydrogenase is an oxidoreductase capable of regenerating both NADH and NADPH depending on the oxidised form of the cofactor available [18-20]. Usually, the glucose dehydrogenase gene is cloned from *Bacillus subtilis or Bacillus* spp. to implement the system in whole cell biocatalysis [20]. The cofactor regeneration process proceeds from conversion of glucose to gluconolactone with the generation of one NADH per turn of the pathway. As such, the pathway is independent of oxygen and could be implemented in anaerobic biotransformation, which is a mode of operation in industrial biocatalysis. This contrast with the more modern NADH oxidase system. In this system, NADH oxidase regenerates one NADH from NAD^+, but requires an input of molecular oxygen for conversion into either H_2O_2 or water [16]. While water is an innocuous product that do not complicate downstream separation, requirement of oxygen by NADH oxidase meant that it could not be applied in anaerobic biotransformation systems. On the other hand, glycerol dehydrogenase has also been shown to offer good NADH regeneration capability through the conversion of glycerol to dihydroxyacetone [14]. Specifically, experiment studies demonstrate a positive correlation between glycerol provision and NADH regeneration [14]. Since the cofactor regeneration system does not require oxygen for function, glycerol dehydrogenase like glucose dehydrogenase could be applied in anaerobic biotransformations. Similarly, formate dehydrogenase that converts formate to carbon dioxide is also insensitive to oxygen, and could be used in an anaerobic biotransformation process [15]. The two-step ethanol utilisation pathway proposed in this work shares similar characteristics, and could be applied in whole cell biocatalysis under anaerobic conditions. Ethanol conversion to acetate has been tapped as a means of regenerating NADPH through the combined actions of alcohol dehydrogenase and aldehyde dehydrogenase [21, 22], but a pathway that connects ethanol to acetyl-CoA as proposed here has not been utilized for NADH regeneration in biocatalysis or metabolic engineering.

From the above discussion, it would appear to the novice in the field of biocatalysis that the process of cofactor regeneration is powered by single enzyme capable of converting reduced form of cofactor to oxidised form. But, more recently, pathways with enzymes capable of regenerating cofactors have also been employed in cofactor regeneration [23-25]. This trend affords the more ready incorporation of cofactor regeneration concepts in metabolic engineering [25], even though pathway-based cofactor regeneration system is also equally applicable in biocatalysis. For example, pathways involving methanol oxidation with cofactor regeneration [23], and use of cofactor regeneration friendly enzymes in xylose metabolization pathway [25] have afforded improved product yield. Compared to single enzyme, expressing a pathway to afford cofactor regeneration capability may incur additional metabolic burden to the cell, but if the pathway could be creatively integrated into cellular metabolism, it may afford opportunities for generation of a useful secondary product in addition to the main biotransformation product.

Ethanol utilization pathway is construed in this project as a linear series of two reactions that guide ethanol first to acetaldehyde and subsequently to acetyl-CoA. Such a pathway architecture has not been utilized in NADH regeneration given that previous attempts on using ethanol for NADPH regeneration centered on the pathway from ethanol to acetate. As designed, the ethanol utilization pathway that ends in acetyl-CoA is not unfamiliar to *E. coli*. Specifically, an endogenous alcohol/aldehyde dehydrogenase in *E. coli*, adhE, also uses identical pathway steps to accomplish ethanol to acetyl-CoA biotransformation [7]. Hence, adhE may be a good candidate to replace the two-gene ethanol utilisation pathway in later parts of the project. Ultimately, choice of the architecture of the ethanol utilisation pathway would depend on the relative performance of the two-gene pathway relative to adhE.

Model reaction chosen for this project is the reduction of ketone to alcohol. Two substrates: one hydrophobic and one hydrophilic would be used as substrates to understand how the ethanol utilisation pathway would perform in both two-phase and single-phase biotransformation systems. Specifically, acetophenone which is hydrophobic would be used in a two-phase biotransformation system, while hydrophilic butanone would be utilized in a single phase biotransformation. The enzyme, cpsADH, chosen for the biotransformation reaction is a carbonyl reductase from *Candida parapsilosis*, which could also serve as an alcohol dehydrogenase. As an oxidoreductase, cpsADH is a redox cofactor dependent enzyme, and need NADH to convert ketone to an alcohol.

Given that the focus of this project is in the evaluation of the relative efficacy of ethanol utilization pathway in regenerating NADH in support of a biotransformation system, a simple and efficacious proven model reaction is desired. Such a model reaction arrives in the form of a ketone to alcohol biotransformation mediated by cpsADH enzyme [26, 27]. The enzyme is a proven player in biocatalysis given its wide substrate scope and high activity in a variety of settings [26-28]. Specifically, it could reduce a variety of aliphatic, aromatic, and cyclic ketones, as well as aldehydes and ketoacetals with high reaction rates [26, 28]. More importantly, it has relatively broad pH optimum between 6.5 and 9.0, and temperature optimum range between 36 and 42°C with good stability [26]. Able to perform biotransformation in both two-phase and single-phase whole cell biotransformation system, cpsADH could also be used in *in vitro* biocatalytic reactions. Moreover, the crystal structure of cpsADH in complex with redox cofactor has been solved (Protein Data Bank), which affords a structural model on which directed evolution approaches could be based, or more fundamentally, understanding the mechanistic basis of biocatalysis reactions mediated by the enzyme.

Choice of ketone to alcohol reduction as model biotransformation reaction comes from the increasingly facile nature in which biocatalytic process could mediate formation of chiral alcohols [29]. But given the focus of this research is in demonstrating the utility of the ethanol utilization pathway in NADH regeneration, chirality of the alcohol product would thus not be assessed. The first ketone substrate for this segment of the work is acetophenone, whose reduced product, 1-phenylethanol, is a useful derivative for the pharmaceutical and fine chemical industry. Given that both acetophenone and 1-phenylethanol are endowed with phenyl rings, they could only be solubilized in organic solvent such as n-hexadecane. This brings forth the need for a two-phase biotransformation system. On the other hand, butanone is chosen as the second substrate given its high water solubility that affords the use of aqueous single phase biotransformation system. The biotransformation product, 2-butanol, is useful as a biofuel as well as a platform chemical for the derivation of other useful product. Both biotransformation reactions described above are well documented reactions [30, 31], and should serve as good model reactions, which help reduce the risk level inherent in this project.

Both the biotransformation and cofactor regeneration reactions could theoretically be carried out using purified enzymes, but high cost and complexity of purifications makes a whole-cell biocatalytic process appealing [32]. More importantly, a whole-cell biocatalysis process protects the enzymes from harsh environments and interfacial stress such as those present in aqueous/organic interfaces in two-phase biotransformation [32]. But, whole-cell biocatalysis is not without challenges, chief of which is mass transfer constraints in bringing the substrate into the cell and the concomitant process of bringing product out of the cell [32]. Overall, high cost of enzyme purification and general stability problem of enzymes in different buffers still recommend the use of whole-cell biocatalysis for biotransformation.

Whole-cell biocatalysis meant that a host cell needs to be chosen for performing the biotransformation with cofactor regeneration. Bacterial cells readily come to mind as candidate host cells given their robustness and resilience to a variety of environmental and nutritional conditions.

Escherichia coli is a commonly used host strain for biotransformation reactions. This likely comes from two reasons: (i) existence of facile genetic tools for implementing heterologous genetic pathways in the host cell for biotransformation and cofactor regenerations, and (ii) extensive knowledge on culturing and experimental use of this biotechnology workhorse. But, beyond these two reasons, relatively little is known about the exact reasons for the choice of *E. coli* as host cells for a variety of biotransformation reactions. One less often noted reason may be the presence of transporters useful for importing and effluxing substrates and products in and out of *E. coli* cells for whole cell biocatalysis. Indeed, transporter engineering has recently taken on importance as seen in a couple of studies detailing how enhancing transport activity help improve product titer [33, 34].

Two modes of operation are common in biocatalysis: resting cells and growing cells. Resting cells refer to a state in a cell exposed to limiting nutrient conditions [15, 35]. Specifically, in the context of biocatalysis, resting cells are typically provided a carbon substrate for powering cofactor regeneration, but no nitrogen substrate. This way, cells could not grow and are assumed to be in a resting state. But, this does not mean that resting cells are not physiologically active in catalyzing the biotransformation of substrate to product with the aid of cofactors regenerated by a cofactor regeneration enzyme. Perhaps, resting cell is a misnomer as these cells are active in biotransformation. Relatively little is known about the gene expression pattern of resting cells, which will be a topic of interest to biocatalysis.

On the other hand, growing cells naturally refer to cells in a growing state [36]. To support cell growth, growing cells are typically cultivated in a full complement growth medium infused with biotransformation substrate and co-substrate. Growing cells are used when a driving force could be set up to force flux down a substrate utilisation pathway to help regenerate more cofactor in support of a biotransformation reaction. For a driving force to develop, the substrate utilisation pathway needs to end at a growth promoting compound that connects to central carbon metabolism. Ethanol utilisation pathway fulfils this criterion given that it ends with acetyl-CoA, an important cellular building block for turning tricarboxylic acid cycle as well as supporting lipids synthesis and gluconeogenesis. Theoretically, in a growing cell biotransformation experiment with ethanol as co-substrate, cells with ethanol utilisation pathway would deplete acetyl-CoA through cell growth processes. Replenishment of acetyl-CoA from ethanol utilisation would then supply the flux for NADH regeneration; thereby, providing more cofactor to support biotransformation and resulting in higher product yield. From a performance perspective comparing resting and growing cells, the latter would outperform resting cells if a driving force could be set up during cell growth.

Overall, cofactor regeneration is an established approach for providing NADH or NADPH cofactors in support of redox-dependent biotransformation reactions in whole cells. Despite the dominance of glucose dehydrogenase system in NADH regeneration, efforts are still being expended in searching for alternative systems for regenerating NADH with higher efficiency and lower substrate cost. Furthermore, there is a movement in the field away from single enzyme cofactor regeneration system towards pathway-based approaches, with the latter offering a route for integration with downstream stream metabolically engineered pathway. Ethanol utilisation pathway as described in this work is not new, given the presence of endogenous adhE enzyme with identical pathway architecture. In the realm of cofactor regeneration, ethanol to acetate conversion pathway powered by an alcohol dehydrogenase in sequence with an aldehyde dehydrogenase has been tapped for NADPH regeneration. What is new in this work is the proposed use of ethanol utilisation pathway linking ethanol to acetyl-CoA for NADH regeneration, and evaluation of its efficacy with respect to glucose dehydrogenase system. Such a research goal help defines the use of an established ketone to alcohol biotransformation mediated by a proven enzyme (cpsADH) to provide a system-level readout of the whole-cell biotransformation system.

4. Materials and Methods

4.1. Chemicals and Medium Composition

LB medium was purchased from Difco and used as is. Composition of LB medium was (g/L): Tryptone, 10.0; NaCl, 10.0; Yeast extract, 5.0. Composition of potassium phosphate buffer was (g/L): K_2HPO_4, 12.54; KH_2PO_4, 2.31. Composition of modified M9 medium was (g/L); K_2HPO_4, 6.8; KH_2PO_4, 3.0; NH_4Cl, 1.0; NaCl; 2.0; Yeast extract, 1.0; Ethanol, 10.0.

Composition of high cell density medium was (g/L): K_2HPO_4, 12.54; KH_2PO_4, 2.31; D-Glucose, 4.0; NH_4Cl, 1.0; NaCl; 5.0; Yeast extract, 12.0; $MgSO_4$, 0.24. Antibiotics used were ampicillin (100 μg/mL) and spectinomycin (50 μg/mL). Acetophenone, 1-Phenylethanol, glucose, butanone, 2-butanol, n-hexadecane and ethyl acetate were of analytical reagent grade (99%) and purchased from Sigma Aldrich.

4.2. Plasmid Construction

Ethanol utilization pathway comprises two genes: *adh2* and *ada* were placed under the control of an autoinducible promoter, P_{thrC3} and inserted into a plasmid backbone with ampicillin as antibiotic resistance marker. Adh2 was cloned from *Saccharomyces cerevisiae* and ada was cloned from *Dickeya zeae* by my labmate, Ms Liang Hong, from whom I inherited the initial pathway. pMB1 serves as the origin of replication in this plasmid. On the other hand, cpsADH alcohol dehydrogenase enzyme was cloned from *Candida parapsilosis* and placed under the control of an autoinducible promoter, P_{thrC3}, and inserted into a plasmid backbone with spectinomycin antibiotic resistance marker. I inherited this plasmid from Mr. Liu Hongyuan from the lab. The replication origin of this plasmid is pA15. Given their differing antibiotic resistance marker and replication origin, both plasmids could be stably maintained in *E. coli* BL21 (DE3). Both plasmids were transformed into *Escherichia coli* BL21 (DE3) using standard heat shock transformation method.

4.3. General Procedure for Cloning of Genes from Bacterial Species

Glucose dehydrogenase (gdh, *Bacillus subtilis*), aldehyde dehydrogenase (aldA, *E. coli*) and alcohol/aldehyde dehydrogenase (adhE, *E. coli*) were cloned from their respective bacterial species. Using appropriate primers, the respective genes were cloned through colony PCR in the thermocycler (Biorad, T-100). Specifically, 1 μL of a 16 hour bacterial culture in LB Miller medium was used as template for PCR amplification. PCR cycling conditions were: 98°C for 5 mins, followed by 34 cycles of (98°C for 8 sec, 55°C for 15 sec, and 72°C for 120 sec), with final extension of 72°C for 5 mins. The PCR product was clean-up through gel electrophoresis (130V for 30 mins, Biorad power pack) followed by gel purification using Thermo Fisher Scientific gel purification kit (GeneJet Gel Extraction Kit, Cat No: #0692). The resulting gene fragment was ligated with the plasmid backbone using a ligase-free ligation method developed in the lab that includes incubation with 10 mM $MgCl_2$ in a thermocycling step in a PCR thermocycler (Biorad, T-100). The cycling programme was as follows: 80°C for 1 min, 68°C for 10 mins, and infinite hold at 4°C. DNA assembly method used was developed in the lab and is known as the GT standard [37].

Supplementary Table 1. Primers used in the ethanol cofactor regeneration project

Gene	Primer Name	Sequence
GDH	Glucose dehydrogenase-Forward	5'-tatccggatttaaaaggaaaagtcgtcgc-3'
GDH	Glucose dehydrogenase-Reverse	5'-accgcggcctgcctggaa-3'
aldA	Aldehyde dehydrogenase-Forward	5'-tcagtacccgttcaacatcctatgtatat-3'
aldA	Aldehyde dehydrogenase-Reverse	5'-ttaagactgtaaataaaccacctgggtct-3'
adhE	Alcohol dehydrogenase-Forward	5'-gctgttactaatgtcgctgaacttaac-3'
adhE	Alcohol dehydrogenase-Reverse	5'-ttaagcggattttttcgcttttttctcag-3'

After plasmid ligation, colony PCR with appropriate primers were used to confirm ligation of gene fragment with plasmid backbone.

Amplified gene fragment was extracted and purified with a gel purification kit (GeneJet Gel Extraction Kit, Cat No: #0692). Sequence of the gene fragment was verified by Sanger sequencing (BioBasic Singapore). Primers used for the experiments are as listed in Supplementary Table 1.

4.4. Cell Cultivation

E. coli was inoculated in 10 mL LB medium with 100 μg/mL ampicillin and 50 μg/mL spectinomycin in 125 mL shake flask. Incubation conditions were 37°C and 225 rpm for all cultivations. The incubator used was N-Biotek NB-205 Incubator-shaker. After 5 hours of cultivation, 0.25 mL of seed culture served as inoculum for 25 mL of LB medium with antibiotics in a 125 mL shake flask and the experiment cultures were cultivated for 20 hours. Optical density of the culture broth was measured at 600 nm with a UV-Visible spectrophotometer (Ambersham Biosciences). Appropriate dilution with deionized water was performed for samples with absorbance that exceed 1. Cells were harvested for biotransformation by centrifugation at 3300g for 5 minutes. Subsequently, cells were resuspended with 89 mM potassium phosphate buffer for a washing step, which included centrifugation at 3300g for 5 minutes. Cells were finally resuspended in 89 mM potassium phosphate buffer with a factor of 2 concentration in optical density for resting cell biotransformation. In growing cells biotransformation, cells were not washed with potassium phosphate buffer but were transferred into either LB Miller medium or M9 ethanol medium with a factor of 0.1 in optical density.

4.5. Acetophenone Biotransformation

Given the poor solubility of acetophenone and 1-Phenylethanol in water, a two-phase system comprising n-hexadecane and potassium phosphate buffer was used in biotransformation. Specifically, either 5 or

20 g/L acetophenone was dissolved in 0.5 mL of nhexadecane, while 0.5 mL of potassium phosphate buffer contained *E. coli* cells and 10 g/L ethanol. The reaction mixture was contained in 2.0 mL HPLC vial and incubated anaerobically at 37°C and 225 rpm for 20 hours. Duplicate experiments were carried out. Cells were precultivated in LB medium for resting cells experiments.

At suitable time-points, HPLC vials were removed from the incubator. The contents were transferred to a 1.5 mL microcentrifuge tube and centrifuged at 20000g for 2 minutes. The upper organic phase containing the substrate and product was aliquoted and diluted with ethyl acetate prior to gas chromatography-mass spectrometry (GC-MS) analysis for detecting acetophenone and 1-Phenylethanol. Following that, the organic phase was carefully removed by pipetting; thereby, leaving behind an aqueous phase where 200 μL was aliquoted, filtered through a 0.22 μm nylon filter prior to high performance liquid chromatography (HPLC) analysis for detecting ethanol and acetate.

The performance measure selected for this segment of the work is yield per unit optical density, and is as defined below:

$$\text{Yield per unit OD} = \frac{\text{Final Phenylethanol conc}}{\text{Initial acetophenone conc} - \text{Final acetophenone conc}} \times \frac{100}{\text{Final optical density}}$$

4.6. Butanone Biotransformation

Since butanone and 2-butanol are highly water soluble, only a single-phase biotransformation system is required. Specifically, 10 g/L of butanone is added to either potassium phosphate buffer (resting cells) or growth medium (growing cells) together with 10 g/L of glucose or ethanol for biotransformation. Concentrations of butanone and ethanol were titrated in experiments aimed at understanding the effect of substrate toxicity effect or whether provision of higher ethanol concentration would drive higher biotransformation efficiency, respectively. The reaction mixture was contained in 2.0 mL HPLC vial and incubated anaerobically at 37°C and 225 rpm for 44 hours. Triplicate experiments were carried out.

At suitable time-points, HPLC vials were removed from the incubator. The contents were transferred to a 1.5 mL microcentrifuge tube and centrifuged at 20000g for 2 minutes. 200 μL of the supernatant was aliquoted, filtered through a 0.22 μm nylon filter prior to high performance liquid chromatography (HPLC) analysis for detecting, butanone, butanol, ethanol and acetate.

The performance measure selected for this segment of the work is yield per unit optical density, and is as defined below:

$$Yield\ per\ unit\ OD = \frac{butanol\ conc}{Initial\ butanone\ conc} \times \frac{100}{Final\ optical\ density}$$

4.7. Gas Chromatography Mass Spectrometry and High Performance Liquid Chromatography Analysis

Gas chromatography mass spectrometry was used in analysing the concentration of acetophenone and 1-Phenylethanol in Agilent 7890 GC and Agilent 5977B MSD. Injection volume used was 1 μL and the oven temperature profile used was 50°C for 1 min, ramp from 50 to 180°C at 10°C/min, followed by another ramp from 180 to 280°C at 50°C/min and a hold at 280°C for 3 minutes. MS detector was turned off at 13 minutes to protect the mass spectrometry detector.

High performance liquid chromatography was conducted with an Agilent 1260 series instrument equipped with a Bio-rad Aminex HPX-87H column. 5 μL of sample was injected into the column and eluted with 5 mM sulphuric acid at 0.7 mL/min. Isocratic elution was used and the column was maintained at 50°C. Analytes were detected by a refractive index detector.

5. Results and Discussion

The first segment of the work involves the expression of two plasmid vectors each encoding enzymes needed for biotransformation. Specifically, plasmid 1 would carry a pathway for converting ethanol to acetyl-CoA with attendant generation of NADH. An autoinducible promoter, pthrC3, would control the ethanol utilization pathway genes on this plasmid. The second plasmid, on the other hand, carries a gene encoding *cpsadh*, an alcohol dehydrogenase needed for biotransformation, which is under the control of an IPTG inducible T7 promoter. Conceptually, *E. coli* would be cultivated aerobically and the genes on the two plasmids would either be automatically induced (ethanol utilization pathway) or induced by IPTG (cpsADH enzyme). After sufficient time for growth and generation of biocatalysts, cells would be harvested and resuspended with in a phosphate buffer without a nitrogen source. This is the biotransformation phase of the experiment. Ethanol would be added to the cells together with the substrate of the reaction.

5.1. Assessing the Metabolic Burden of Expressing Heterologous Genes in *E. coli* through Growth Assay

Experiment work started with the creation of a strain carrying the ethanol utilization pathway and the cpsADH alcohol dehydrogenase necessary for biotransformation. Both pathways were encoded in separate plasmids with different antibiotic selection markers. The ethanol utilization pathway is controlled by autoinducible promoter pthrC3, and the cpsADH enzyme was under the control of a T7 promoter. Transformation work was done to incorporate both plasmids into the cells of *Escherichia coli* T7 express and the strain was named *E. coli* ethcps1. Subsequent to transformation, growth experiments were carried out to assess the relative fitness of the engineered *E. coli* ethcps1 in different growth media. Given that expression of EUP is under the control of an autoinducible promoter, EUP pathway genes would be expressed during exponential growth, and the extent of its metabolic burden could be assessed via a cell growth assay. Experimental results revealed that engineered *E. coli* ethcps1 grew well in LB Miller and high cell density (HCD) medium with the HCD medium offering a higher maximal optical density of 8.9 compared to 3.5 in LB Miller medium (Figure 11). Such maximal optical density values highlight that autoinducible expression of EUP pathway gene did not result in significant metabolic burden and hamper cell growth. However, the maximal optical density in HCD medium was reached in 48 hours which would add to the time needed for the biotransformation workflow.

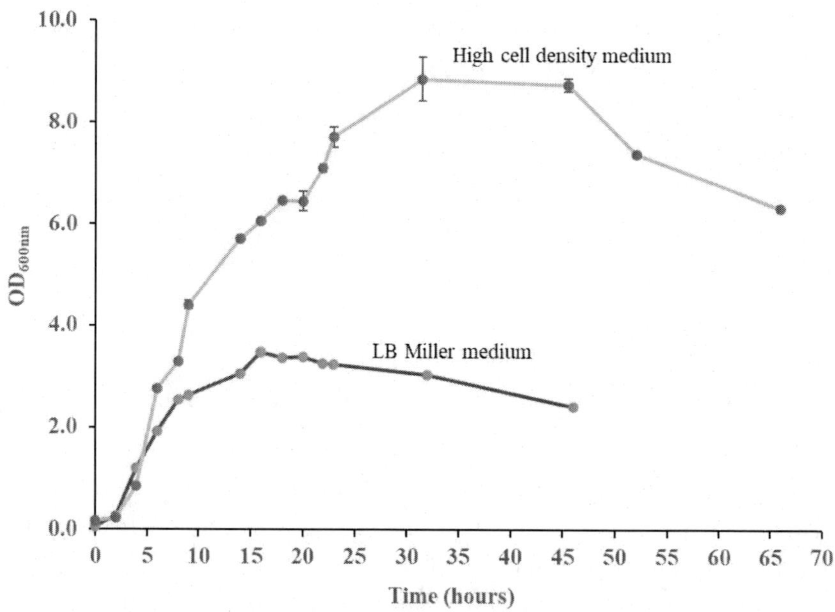

Figure 11. Growth profile of *E. coli* ethcps1 in different growth medium, illustrating the relative lack of metabolic burden from the autoinduced expression of ethanol utilization genes.

5.2. Evaluating the Need for an Organic Phase to Dissolve Hydrophobic Acetophenone

The substrate of the biotransformation reaction, acetophenone, would need to be dissolved in an aqueous medium to allow a greater bioavailability to *E. coli* ethcps1 cells. However, acetophenone and the product, 1-Phenylethanol, are only sparingly soluble in water. Because acetophenone and 1-Phenylethanol are highly soluble in acetonitrile, and acetonitrile and water being miscible with each other, acetonitrile is a possible co-solvent for improving the solubility of acetophenone and 1-Phenylethanol in water. However, potential toxicity of high concentration of acetonitrile on cells remain unknown. Thus, experiments were conducted to evaluate the toxicity effect of acetonitrile on *E. coli* cells.

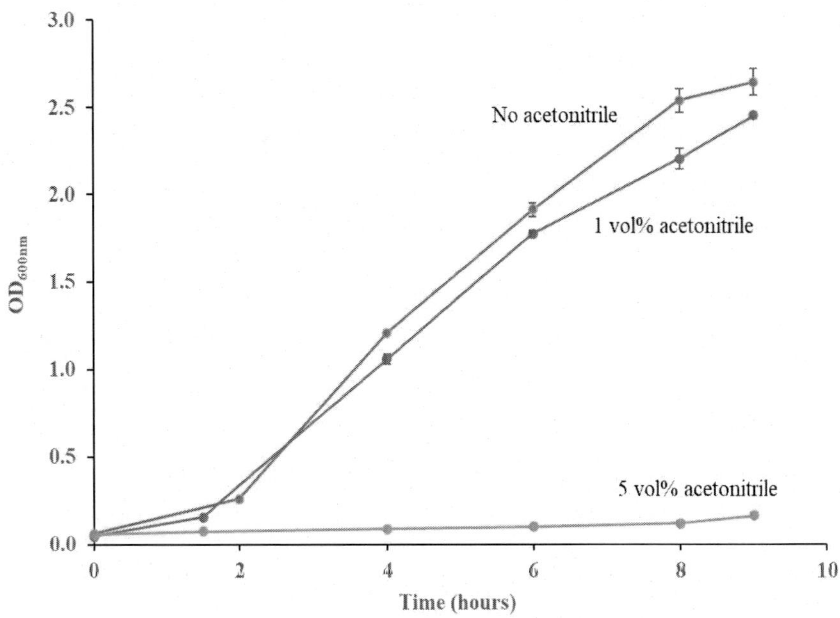

Figure 12. Examining the toxicity effect on growth of a potential co-solvent, acetonitrile, for enabling better solubility of the biotransformation substrate, acetophenone, in the aqueous phase.

Experimental data revealed that acetonitrile exerted a concentration-dependent toxicity effect on cell growth with 5 vol% acetonitrile rendering the cells in a non-growing state (Figure 12). On the other hand, 1 vol% acetonitrile was well tolerated by the *E. coli* cells with cell growth profile similar to those not exposed to acetonitrile. However, there was some toxicity effect induced by 1 vol% acetonitrile that resulted in a slightly smaller optical density obtained. Overall, 1 vol% acetonitrile could be used as a co-solvent for improving the solubility of acetophenone without inducing significant toxicity effect on the cells. However, the amount of acetophenone that could be brought into solution with 1 vol% acetonitrile is small, and attention in this initial phase of the project switched to the use of an organic phase, n-hexadecane, to aid the dissolution of acetophenone.

5.3. Selecting Suitable Promoters for Controlling Cpsadh Expression

Inducer is commonly used to induce the expression of heterologous genes in constructed pathways. In this project, IPTG is used as inducer for the cpsADH enzyme under the control of a T7 promoter on pET28a plasmid. However, IPTG could exert metabolic burden on cells that manifest as reduced biomass formation. Indeed, experimental data from *E. coli* T7 express cells with two plasmids exposed to different concentrations of IPTG revealed that IPTG exerted metabolic burden on cells. Specifically, the maximal optical density attained by cells exposed to 0.5 mM IPTG was lower than that of those exposed to 0.1 mM IPTG. In addition, provision of IPTG resulted in significant metabolic stress and burden that resulted in a smaller maximal optical density of 2.0 to be obtained during growth of the bacterium in LB Miller medium (Figure 13). This is in contrast to the maximal optical density of about 3.4 obtained in engineered *E. coli* cells not exposed to IPTG. Importantly, amongst all tested IPTG concentrations, cell growth rate was not retarded. However, IPTG exerted its effect on maximal optical density obtained, which reveals that metabolic burden resultant from over-expression of cpsADH may be the key reason underlying poorer growth of bacterium. This further confirmed that observed reduced optical density was not resultant from toxicity of IPTG. In general, lower IPTG concentration was correlated with higher maximal optical density of engineered *E. coli*, but there was a substantial reduction in maximal optical density even at the lowest tested concentration of IPTG at 0.1 mM. Because cell density is important for assuring a higher concentration of enzymes for fast biotransformation, a lower concentration of IPTG would need to be used for improving the cell density obtained. However, the overall results suggest that an IPTG inducible promoter may not be suitable for controlling the cpsADH gene in the second plasmid due to its significant metabolic burden effect.

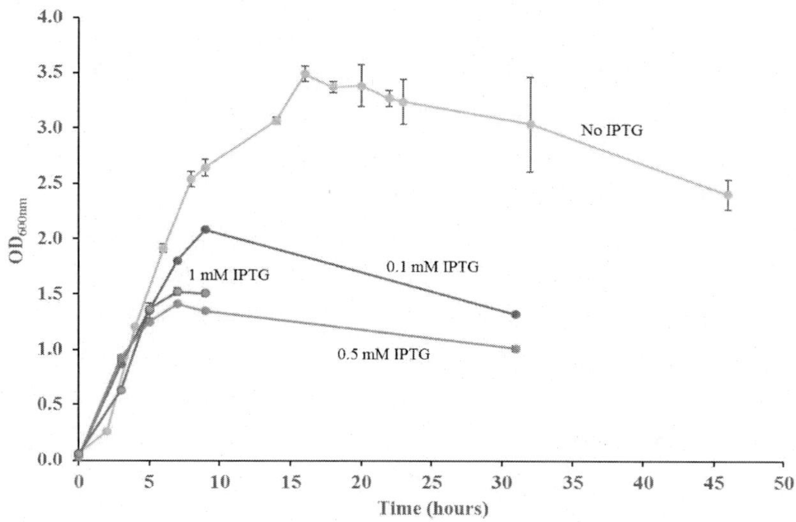

Figure 13. Effect of different IPTG concentration on cell growth performance as measure of metabolic burden induced by IPTG.

As inducer necessary for inducing the expression of heterologous genes are expensive, efforts were directed to place the cpsADH enzyme under the control of an autoinducible promoter. This comes in addition to preliminary experimental work revealing the significant metabolic burden induced by IPTG addition. A new plasmid with cpsADH enzyme under the control of an autoinducible promoter, pthrC3 was constructed and transformed into *E. coli* BL21 (DE3) together with another plasmid that encode the genes for ethanol utilization, which are also under the control of autoinducible promoter, pthrC3. This new strain, *E. coli* BL21 (DE3) ethcps2 will be used for downstream biotransformation experiments.

5.4. Growth Performance of *E. coli* Ethcps2 Relative to *E. coli* Ethcps1

Prior to commencing biotransformation experiments that evaluate the efficacy of the ethanol utilization pathway on NADH regeneration efficacy, it is important to first gain an understanding of the effect of a switch from a T7 promoter to pthrC3 autoinducible promoter for controlling cpsADH expression on metabolic burden in the cell. This comes about due to the possibility that autoinducible promoter such as pthrC3 might induce a high expression level for heterologous genes that subsequently result in high metabolic burden.

To check the metabolic burden resultant from a switch in promoter for cpsADH enzyme, growth experiments in LB Miller experiment were carried out under aerobic conditions at 37°C and 225 rpm shaking in 125 mL shake flasks. Experiment results revealed in Figure 14 indicated that both *E. coli* ethcps1 (T7 promoter for cpsADH) and *E. coli* ethcps2 (pthrC3 promoter for cpsADH) achieved similar optical density during culture in LB Miller broth. However, the initial growth rate of *E. coli* ethcps2 was slightly faster than that of *E. coli* ethcps1; thereby, indicating that the autoinducible promoter did not engender a high expression level of heterologous genes that could detrimentally affect the growth of the bacterium. On the other hand, the results also suggest that the new plasmid encoding cpsADH in *E. coli* ethcps2 does not need higher level of maintenance (replication and expression) than the pET28a plasmid that carried cpsADH enzyme in *E. coli* ethcps1. Hence, in conclusion, use of the pthrC3 autoinducible promoter for controlling cpsADH expression is feasible, and would not induce a high metabolic burden that resulted in reduced growth rate.

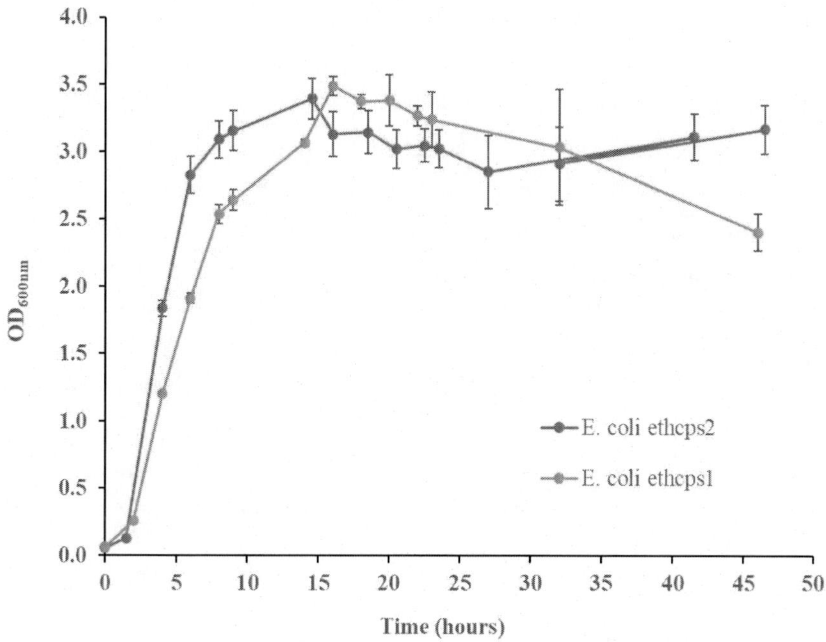

Figure 14. Growth performance of *E. coli* ethcps2 relative to *E. coli* ethcps1 in LB medium for assessing the metabolic burden of replicating constituent plasmids of the two engineered strains.

5.5. Use of a Two-Phase System for Acetophenone Biotransformation for Evaluating Efficacy of Ethanol Utilization Pathway Powered NADH Regeneration

As the solubility of acetophenone in aqueous phase was low even with the help of a cosolvent, acetonitrile, the research focus of the project shifted to the use of a two-phase system for biotransformation. Specifically, n-hexadecane was chosen as the organic phase to allow a higher solubility of both the substrate and product in the biotransformation system. In particular, provision of a higher concentration of substrate acetophenone would hopefully help improve the production titer of the product, 1-Phenylethanol by shifting the equilibrium position towards product formation. However, bioavailability of the substrate, acetophenone, to the cells remain a potential problem, and to this end, good and vigorous mixing was necessary for the biotransformation to occur. Concentrations of acetophenone and 1Phenylethanol were analysed by GC-MS, while that of co-substrate ethanol and metabolic byproduct, acetate, in the aqueous phase were analysed by HPLC refractive index detector.

Preliminary experiments revealed that the two-phase biotransformation system afforded the conversion of acetophenone to small amounts of 1-Phenylethanol with input acetophenone concentration of 5 g/L and 10 g/L. Identity of 1-Phenylethanol product was confirmed with GC-MS through comparison of detected peak with that of a library (Figure 15a). This suggests that the entire biotransformation system is functional given that the EUP pathway must be able to regenerate NADH in order to allow biotransformation of acetophenone to occur in resting cells. Although the titer of 1-Phenylethanol was not high, it nevertheless demonstrates the feasibility and utility of the two-phase transformation in enabling cofactor regeneration powered acetophenone biotransformation. Further experiments aimed at examining the range of concentrations suitable for recombinant *E. coli* BL21 (DE3) ethcps2 to mediate biotransformation of acetophenone to 1-Phenylethanol revealed a positive correlation between acetophenone concentration and product titer with the acetophenone concentration range of 1 and 20 g/L. Specifically, higher conversion of acetophenone was observed at 20 g/L compared to 1 g/L probably due to the reversible nature of the reaction where a higher substrate concentration would drive the equilibrium position forward (Figure 15b). Finally, attempts were also made to increase the cell density of the biotransformation system and results revealed that a higher cell density enabled a higher conversion of acetophenone, which dovetails with theoretical predictions that higher cell numbers present higher enzyme concentration for mediating biotransformation and allowing higher product titer to be achieved.

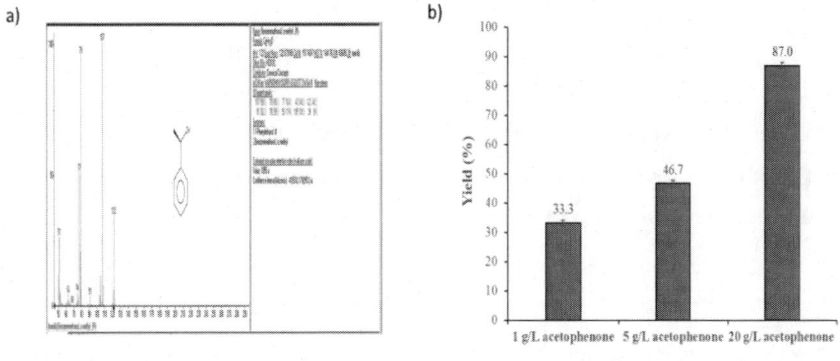

Figure 15. Summary of preliminary experiments in acetophenone biotransformation powered by EUP NADH regeneration. a) Confirmed detection of 1-Phenylethanol product by GC-MS. b) Effect of initial acetophenone concentration on yield of 1-Phenylethanol product.

Control experiments, on the other hand, revealed that heterologous ethanol utilization pathway and cpsADH were needed for the biotransformation of acetophenone. However, recombinant *E. coli* BL21 (DE3) cps2 with cpsADH enzyme but without ethanol utilization pathway could mediate substantial conversion of acetophenone when fed with 10 g/L of ethanol. This revealed that the endogenous alcohol dehydrogenase (*adhE*) that metabolize ethanol could generate substantial amount of NADH to support biotransformation. Given the observation that ethanol could be converted by adhE for generating NADH, a control experiment where there was no provision of ethanol was designed. Results revealed that there was no conversion of acetophenone under the scenario of no ethanol provision, which meant that there was insufficient amount of NADH in the resting cells.

E. coli possess endogenous enzymes capable of utilizing glucose and ethanol. What is relatively unknown is the capability of these enzymes to regenerate NADH and their impact on acetophenone biotransformation. Using *E. coli* strain without an ethanol utilization pathway but with cpsADH enzyme (a strain named as *E. coli* cps2), an attempt was made in understanding whether glucose could be metabolized by the cells to generate sufficient NADH to support the biotransformation of acetophenone. Experiment results with resting cells revealed that substantial conversion of acetophenone to 1-Phenylethanol occurred, and was slightly lower than that achieved with the metabolization of ethanol by adhE in *E. coli* cps2 (Figure 16). More importantly, the effect was reproducible over two concentrations of acetophenone. This suggests that endogenous glucose metabolism is less efficient in regenerating NADH compared to adhE enzyme.

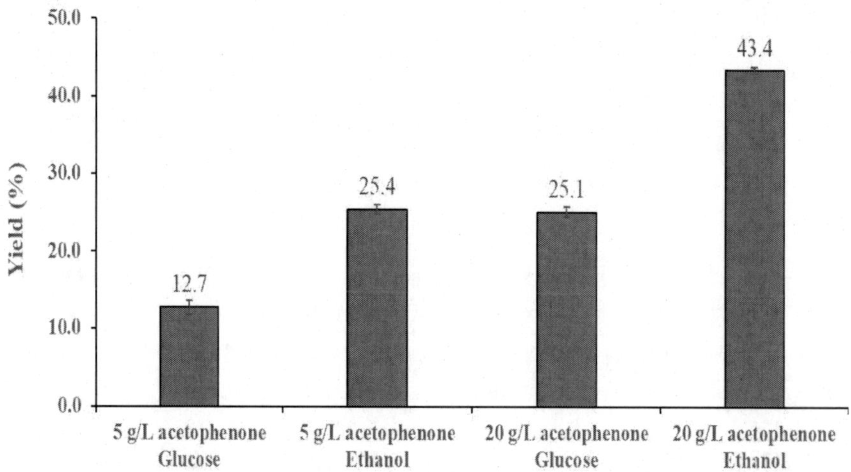

Figure 16. Assessing the ability of endogenous glucose and ethanol utilization pathways in providing NADH in support of acetophenone biotransformation.

5.6. Optimization of Acetophenone Biotransformation Process with Ethanol Powered NADH Regeneration

Because conversion of acetophenone could be limited by thermodynamic equilibrium, provision of more acetophenone might be able to move the equilibrium position forward. Experiment results revealed that increase in acetophenone concentration was positively correlated with higher conversion. Specifically, conversion of acetophenone ranged between 46.5 and 55.1% for initial acetophenone concentration of 5 g/L, but conversion increased to between 90.5 and 95.4% for initial acetophenone concentration of 20 g/L. However, further increase in acetophenone concentration to 50 g/L did not result in higher conversion (Figure 17). Thus, there is a limit to the concentration of acetophenone that could be used to drive the reaction forward. Another parameter that could drive higher biotransformation efficiency and product yield is the concentration of ethanol provided given that higher concentration of ethanol should lead to greater NADH regeneration and higher 1-Phenylethanol yield. Thus, amount of ethanol provided was also varied to understand the effect of enhanced provision of ethanol on conversion of acetophenone (Figure 17). Results indicated that greater provision of ethanol did result in a slight increase in conversion rates, which corroborated theoretical predictions that higher ethanol concentration would drive EUP that results in greater NADH generation. But, most of the ethanol provided were not utilized which meant that ethanol provision

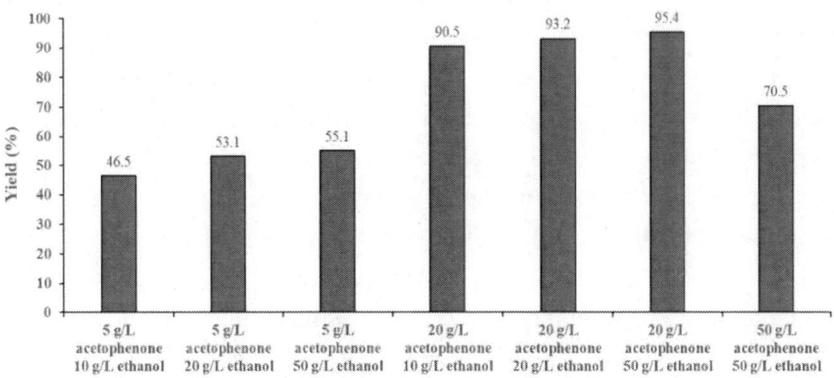

Figure 17. Effect of varying initial acetophenone concentration and ethanol concentration on acetophenone biotransformation efficiency.

is not a limiting factor in this system. Rather, availability of acetophenone in the aqueous phase is a more important factor driving the biotransformation process. Finally, experiments were conducted to assess the toxicity limit of acetophenone and ethanol at 50 g/L each. Observations of lowered biotransformation efficiency and yield suggest that 50 g/L ethanol exerted toxicity effect on the cells that reduced yield in resting cell biotransformation. Given the poor utilization of ethanol at all concentration ranges, the ethanol concentration chosen for future experiment was 10 g/L, while that chosen for acetophenone was 20 g/L due to higher yield at this concentration.

5.7. Using Kinetic Experiments as Tool for Probing and Tuning Acetophenone Biotransformation Dynamics

To help understand the conversion kinetics of acetophenone to 1-Phenylethanol that impact on design of fermentation systems for synthesizing the product, time profile experiments were conducted to obtain kinetic conversion data. At initial acetophenone concentration of 20 g/L and 10 g/L ethanol, uptake of acetophenone was observed to be rapid and decline gradually. However, formation of 1-Phenylethanol was slow and continued till the end of the biotransformation at 23 hours. Thus, the biotransformation seems to be reactionlimited. Ethanol uptake was also rapid and largely completed within 2 hours, but formation of the metabolic byproduct, acetate, was slow and show a similar time profile as 1-Phenylethanol (Figure 18). Overall, 1-Phenylethanol titer was 2.2 g/L, and the yield per unit optical density on acetophenone was 14% (yield on acetophenone was 87.5%).

To reduce fermentation temperature and thus increase process efficiency, the study next moved on to investigate the effect of temperature on biotransformation efficiency. Experimental results revealed that the optimum temperature for biotransformation was 37°C where the yield per unit OD obtained was 14% (Figure 19a). On the other hand, reduced yield per unit OD of about 7-8% was observed at incubation temperatures of 25 and 30°C. Probing the kinetics of product formation reveals that higher temperature afforded higher product titer in the same amount of time (Figure 19b). Although this suggests that enzyme kinetics of cpsADH might be faster at elevated temperature, the reality check is that the biotransformation might be diffusion limited. This comes about through the observation of similar kinetic profile across all three investigated

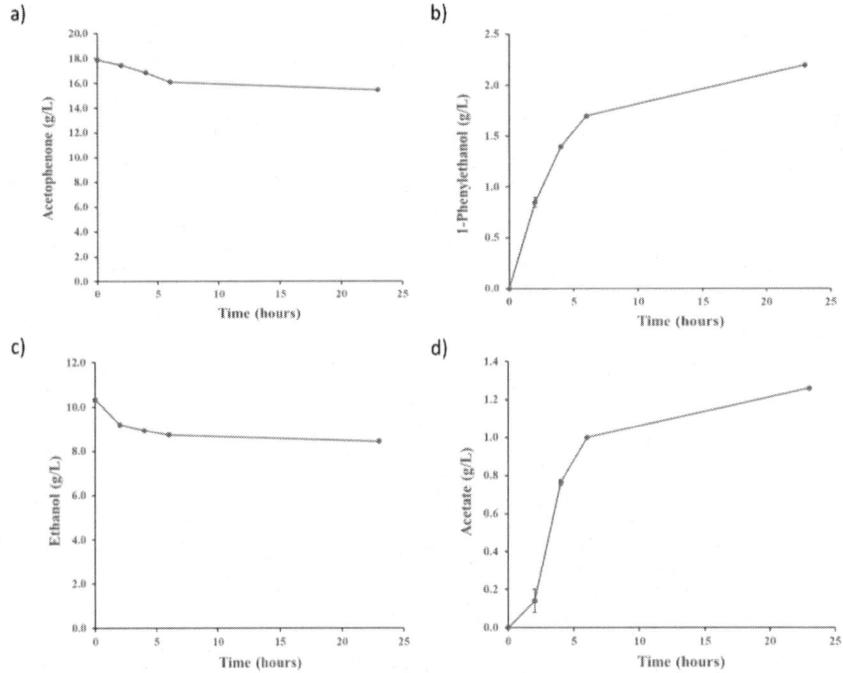

Figure 18. Concentration profiles of different metabolites during acetophenone biotransformation by *E. coli* ethcps2 with ethanol utilization enabled NADH regeneration. a) acetophenone, b) 1-Phenylethanol, c) ethanol, d) acetate.

temperatures that only differ in the amount of product at specified time points. Similar enzyme kinetics across the chosen temperature is the only explanation for the observed similar slopes of the kinetic profiles, which leaves temperature-dependent diffusion of substrate in a two-phase system as the main factor affecting product formation kinetics. Given that similar product titer of 2.0 g/L was obtained at three temperatures used, it indicated that amount of acetophenone uptake played an important role in final yield calculation. Specifically, uptake of substrate acetophenone was observed to be lower at 37°C compared to 25 and 30°C (Figure 19c). This suggests that affinity of *E. coli* transporter for acetophenone was reduced at a higher incubation temperature of 37°C, which opens up opportunities for future engineering of *E. coli* protein transporter for substrate uptake into the cell. On the other hand, higher biotransformation efficiency achieved at 37°C could have been due to higher ethanol uptake at the elevated temperature.

However, comparison of ethanol uptake across the temperature range tested revealed a negative correlation where higher temperature led to poorer ethanol uptake (Figure 19d). Specifically, ethanol uptake declined from 2.30 g/L at 25°C to 1.90 g/L at 37°C. Decline in ethanol uptake with temperature highlights a fundamental limit to biotransformation efficiency and NADH regeneration. Furthermore, the result also suggests sensitivity to ethanol in *E. coli* may be temperature dependent.

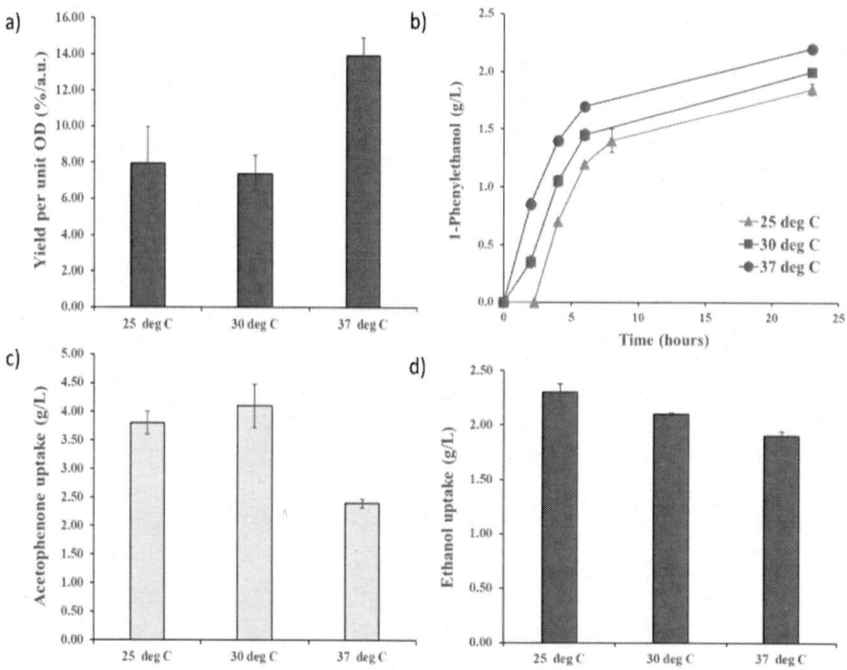

Figure 19. Effect of temperature on acetophenone biotransformation. a) Yield per unit optical density was highest at 37°C, b) time evolution of 1-Phenylethanol concentration also showed that 37°C was the optimal temperature for biotransformation, c) acetophenone uptake at 37°C was noticeably lower than those at 25 and 30°C that points to avenues for transporter engineering, and d) ethanol uptake declined with temperature, which place a fundamental limit on biotransformation efficiency.

As biotransformation of acetophenone to 1-Phenylethanol by cpsADH enzyme is a reversible reaction, a higher concentration of substrate should theoretically move the equilibrium position towards product formation and results in a higher product titer. Thus, the effect of different initial substrate concentration on biotransformation efficiency was investigated in greater detail through kinetic profiling. Experimental results generally concurred with theoretical predictions. Specifically, it was observed that at higher initial substrate concentration, higher yield and product titer was observed. For example, a conversion of 92% and product titer of 2.2 g/L was obtained at 20 g/L initial acetophenone concentration. Conversion decreased to 46.7% and 33.3% when initial substrate concentration was reduced to 5 g/L and 1 g/L respectively (Figure 20a). Conversion, in this case, is defined as the percentage ratio of the amount of 1-Phenylethanol produced relative to amount of acetophenone uptake by the cells. While not all acetophenone that has been uptake into the cells would have been reacted, conversion nevertheless represents a good approximation for biotransformation efficiency for this instance. Thus, experimental results concurred with theoretical predictions based on equilibrium considerations in the case of acetophenone biotransformation.

Higher acetophenone substrate concentration also has a knock-on effect on the kinetics of the process. For example, rate of formation of 1-Phenylethanol was fastest when the initial acetophenone concentration was 20 g/L. This rate progressively reduces with the titration of acetophenone downwards to 5 g/L and 1 g/L (Figure 20b). This indicates that higher acetophenone concentration not only improve conversion and product yield, but also has positive effect with increasing reaction rate, which dovetails with theoretical predictions. Given that the current biotransformation system employs cofactor regeneration, it is also interesting to check if increased acetophenone concentration impacts on ethanol utilization. Results revealed a weak positive dependence between acetophenone concentration and ethanol utilization (Figure 20c). This comes about probably through a driving force enacted by higher consumption of acetophenone that created a demand for higher NADH pool, which could be supplied by higher ethanol utilization in the EUP pathway. Finally, there was a direct correspondence between the 1-Phenylethanol concentration profile and acetate formation profile indicating tight coupling between the biotransformation reaction and EUP pathway (Figure 20d). In being a downstream product from acetyl-CoA, formation and extracellular secretion of acetate could be used as a diagnostic tool for the functional status of the EUP pathway, as well as understanding how carbon flux flow in the system. Given the lack of growth processes in resting cells, acetyl-CoA produced would most likely be channeled to acetate for extracellular secretion, leading to an overall increase in detected acetate concentration.

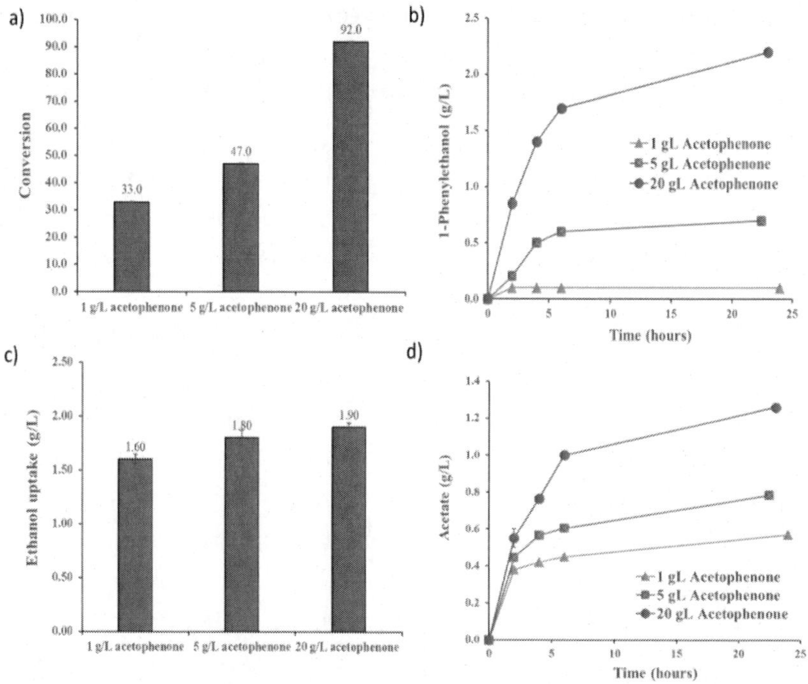

Figure 20. Effect of acetophenone concentration on biotransformation efficiency. a) Conversion was positively correlated with initial acetophenone concentration, b) Higher acetophenone concentration increases the kinetics of 1-Phenylethanol formation, c) Weak dependence between ethanol uptake and acetophenone concentration was observed, d) acetate formation kinetics show correspondence with amount of ethanol uptake, thereby, demonstrating the importance of acetate as a metabolite for monitoring EUP pathway performance.

Work on the project continued with understanding how promoter affects the biotransformation efficiency of acetophenone to 1-Phenylethanol through its effect on the amount of cpsADH enzyme synthesized. Specifically, this segment of the work utilizes two strains: *E. coli* ethcps1 and *E. coli* ethcps2. In *E. coli* ethcps1, cpsADH expression is under the control of an IPTG inducible T7 promoter, while the same gene is under the control of pthrC3 autoinducible promoter in *E. coli* ethcps2. In time profile experiments with 20 g/L acetophenone initial concentration, *E. coli* ethcps1 could only achieve a yield per unit OD of 9.86% compared to 14.66% in *E. coli* ethcps2 where 0.1 mM IPTG was used to induce expression of cpsADH in *E. coli* ethcps1 during the cultivation phase (Figure 21a). 1Phenylethanol titer obtained with *E. coli* ethcps1 was 1.4 g/L compared to 2.2 g/L by *E. coli* ethcps2, which has both ethanol utilization pathway and cpsADH enzyme under the control of auto-inducible pthrC3 promoters (Figure 21b). This suggests that auto-inducible promoters may result in a higher expression level of cpsADH enzyme over T7 promoter; thereby, resulting in a higher yield and product titer. But, what is more intriguing is the present of a lag phase in 1-Phenylethanol formation in resting cells with T7 promoter for cpsADH expression. One possible explanation is that the half-life of cpsADH enzyme may be short; thereby, requiring activation of gene expression in resting cells. However, IPTG was only added during cell cultivation prior to resting cell biotransformation. Thus, what accounts for the observed 1Phenylethanol synthesis should likely be leaky expression of cpsADH from the T7 promoter that yields a small concentration of enzyme that resultant in a smaller yield for the biotransformation product.

Looking at the ethanol uptake parameter for the same set of experiments yields a picture in which ethanol utilization was independent of the cellular dynamics revolving cpsADH mediated biotransformation. Specifically, similar uptake levels were observed for ethanol by cells with T7 or pthrC3 controlling cpsADH expression (Figure 21c). However, ethanol uptake levels may not directly correlate with ethanol conversion levels. This becomes apparent when one looks at Figure 21d which details the concentration time profile for acetate formation. In this plot, acetate formation was faster and higher in cells with pthrC3 promoter compared to cells with T7 promoter. The results point to two important conclusions. First, in cells where there is no formation of 1-Phenylethanol such as during the lag phase in Figure 21b, ethanol was continuously converted to acetyl-CoA and on to acetate which is secreted. Such a scenario is encouraging given that the EUP pathway could function independently from biotransformation even though NADH pool serves as the driving force connecting the two reaction pathways. Second, higher NADH demand created by acetophenone conversion to 1Phenylethanol serves as the key link that propels ethanol conversion to acetyl-CoA and acetate.

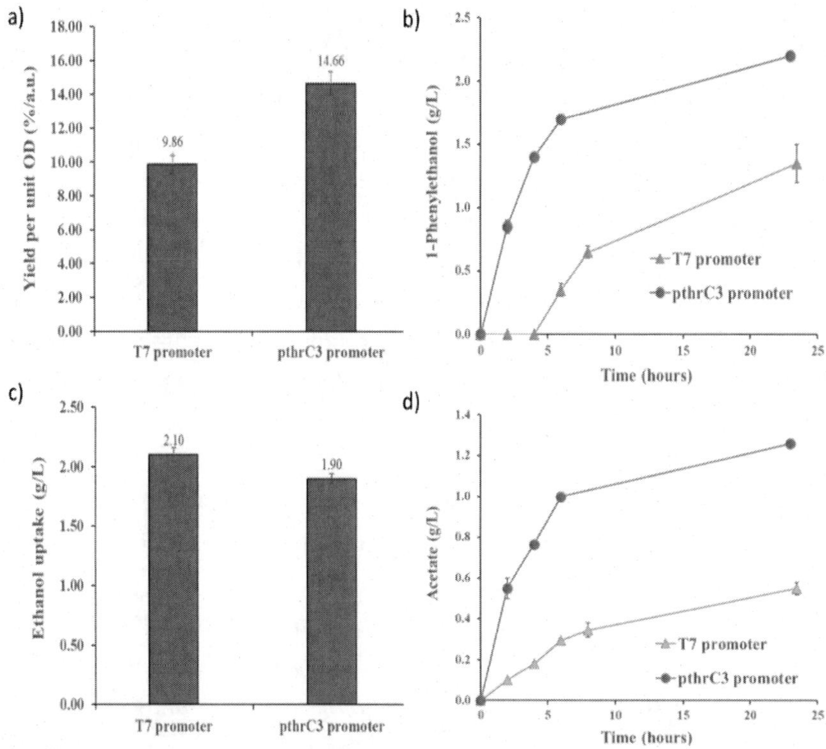

Figure 21. Effect of different promoters for cpsADH expression on biotransformation efficiency. a) Cells with pthrC3 promoter for cpsADH delivered higher yield per unit optical density compared to cells with T7 promoter, b) Delayed formation of 1-Phenylethanol comes from leaky expression of cpsADH genes in resting cells with T7 promoter, c) Similar ethanol uptake was observed in cells with T7 promoter or pthrC3 promoter for cpsADH expression, d) acetate formation profile shows that acetophenone biotransformation is needed to drive ethanol utilization and acetate formation.

In conclusion, recombinant *E. coli* ethcps2 expressing an ethanol utilization pathway and cpsADH alcohol dehydrogenase was constructed and shown to be useful for the biotransformation of acetophenone to 1-Phenylethanol in a two-phase system. Different parameters important to biotransformation efficiency such as temperature, substrate concentration and promoter choice for expressing cpsADH were investigated. Specifically, the optimum temperature for biotransformation was found to be 37°C, while a higher substrate concentration helped improve biotransformation efficiency by driving the equilibrium position of the reversible reaction towards product formation. With regards to tuning the expression level of cpsADH to afford a larger quantity of enzymes for enabling faster biotransformation with better efficiency, experiment tests confirm that pthrC3 autoinducible promoter is preferable compared to T7 promoter in enabling higher acetophenone biotransformation efficiency to be obtained. This comes about likely due to the higher amount of cpsADH synthesized under the control of pthrC3 compared to T7 promoter. Future work in this segment will explore how growing cells perform in acetophenone biotransformation. But first, the capability of the ethanol utilization pathway in enabling growth of *E. coli* in aerobic and anaerobic conditions would need to be assessed.

5.8. Assessing the Functionality of the Ethanol Utilization Pathway in Promoting Growth

To understand whether the ethanol utilization pathway was functional and could be coupled to growth, growth experiments were conducted where ethanol and ammonium chloride were supplemented into the culture broth after the engineered *E. coli* culture had reached stationary phase. Experiment results revealed that a higher optical density was obtained in cultures with supplemented ethanol (5 g/L) and ones with supplemented ethanol (5 g/L) and ammonium chloride (1 g/L) compared to the base case where there was no supplementation.

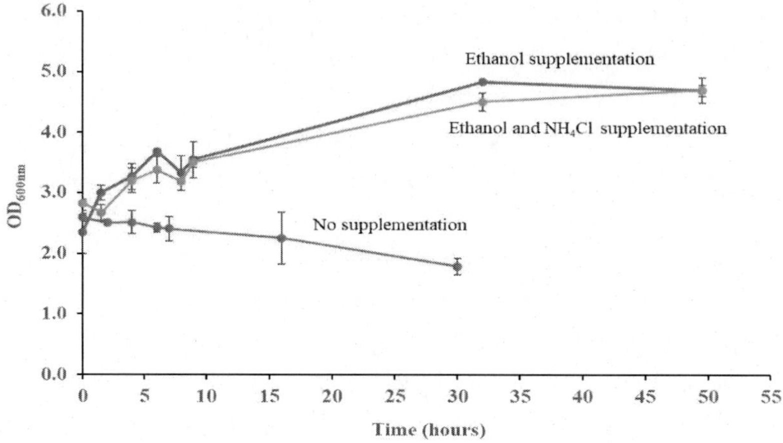

Figure 22. Growth profile of *E. coli* ethcps1 supplemented with either ethanol or ethanol and NH_4Cl 16 hours into the cultivation. Data showed *E. coli* ethcps1 with EUP pathway could grow on ethanol, which suggests that the EUP pathway is functional.

Maximal optical densities obtained were between 4.7 and 4.8 for cultures with supplemented ethanol and ammonium chloride while that for cultures with no supplementation was 3.4 (Figure 22). This indicated that there was growth associated with the use of ethanol as a carbon source by *E. coli* BL21. Another important finding from the experiment was that the engineered *E. coli* could grow to a higher optical density in LB Miller medium supplemented with 5 g/L ethanol without the provision of additional ammonium chloride, which indicated that LB Miller medium is not deficient in nitrogen sources. In essence, LB Miller medium is a nitrogen replete and carbon limited medium.

Besides demonstrating that *E. coli* with EUP pathway and cpsADH could achieve higher optical density value in LB medium supplemented with ethanol at the stationary phase, it is also of importance to further test the EUP growth promoting capabilities in basal medium supplemented with ethanol. To this end, an experiment was set up to observe the aerobic growth performance of *E. coli* ethcps2 with EUP and cpsADH under the transcriptional control of pthrC autoinducible promoter in M9 medium with 10 g/L ethanol. Results revealed that *E. coli* ethcps2 could grow to appreciable cell density of 1.68 in M9 medium supplemented with ethanol at 37°C and 225 rpm in shake flask (Figure 23). Ethanol utilization profile indicated that ethanol utilization started 2 hours into the cultivation with a constant utilization rate. This is correlated with rapid growth of the cells, and affirms the theoretical prediction that ethanol conversion via EUP could generate acetyl-CoA that feeds into biomass formation and growth. During the growth profile, substantial amount of ethanol could be consumed, which suggests that a bottleneck does not exist in ethanol utilization under aerobic conditions.

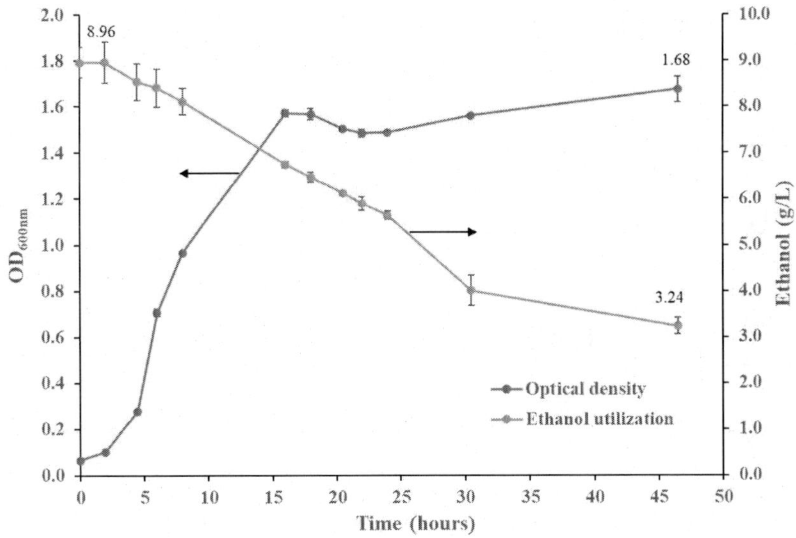

Figure 23. Growth performance of *E. coli* ethcps2 with EUP pathway and cpsADH in M9 + 10 g/L ethanol medium, showing correlated growth and ethanol utilization that suggests that the EUP pathway could promote growth.

While aerobic cell growth assay has demonstrated the utility of the EUP pathway in promoting growth in *E. coli* using ethanol as the main carbon source, it is also important to assess the endogenous capability of *E. coli* to subsist on ethanol. To this end, a strain of *E. coli* with cpsADH but without EUP (named *E. coli* cps2) was utilized. Experiment growth assay involving comparison of growth performance of *E. coli* ethcps2 (with EUP) and *E. coli* cps2 (without EUP) in M9 medium with glucose or ethanol as main carbon source. Results revealed that *E. coli* cps2 could use endogenous ethanol utilization pathway to grow in M9 ethanol medium, but the maximal optical density achieved was significantly less than that obtained by *E. coli* ethcps2 with EUP pathway in the same medium. This indicates that the endogenous ethanol utilization pathway in *E. coli* could not deliver substantial growth to the bacterium with ethanol as main carbon source. Such observations reaffirm that the EUP pathway in *E. coli* ethcps2 is functional and able to contribute to cell growth on ethanol. However, maximal optical density achieved by *E. coli* ethcps2 in M9 ethanol remained lower than that obtained by *E. coli* cps2 in M9 glucose medium, thereby, indicating that *E. coli* endogenous glucose metabolization pathway is more efficient in delivering growth to the organism. Nevertheless, a surprising finding from the data concerns the similar initial growth rate observed during growth of *E. coli* cps2 in M9 glucose and *E. coli* ethcps2 in M9 ethanol medium. Such data suggests that combination of EUP and endogenous ethanol utilization pathway have good kinetic properties at metabolizing ethanol.

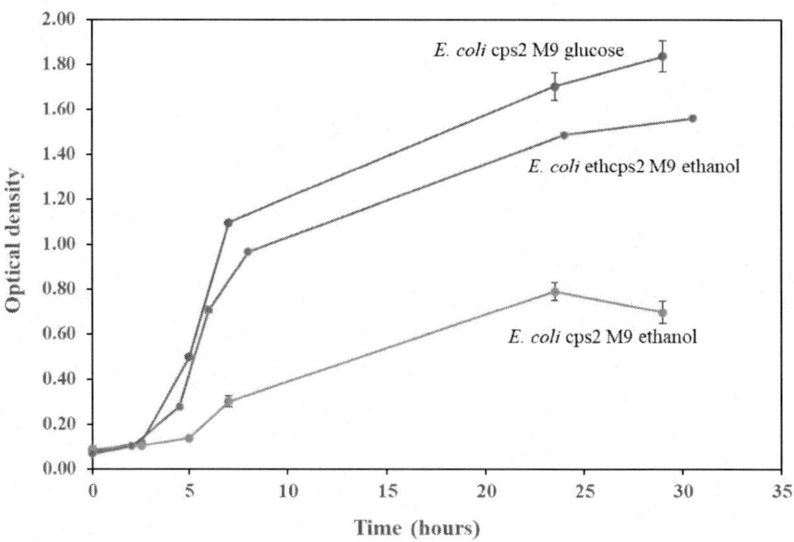

Figure 24. Comparison of growth potential in *E. coli* strains with and without EUP pathway in M9 medium with either glucose or ethanol as main carbon source. *E. coli* cps2 utilizes endogenous glucose and ethanol metabolism pathways, while *E. coli* ethcps2 has an added EUP pathway on top of endogenous enzymes.

The experiments described above pertains to aerobic growth behaviour of strains of engineered *E. coli* with and without EUP pathway. But, given the use of anaerobic conditions for assessing the cofactor regeneration capability of EUP pathway in this project, it is also of importance to examine the anaerobic growth potential of *E. coli* strains with and without EUP pathway. In particular, it is critical to probe the growth potential that could be derived from endogenous utilization of ethanol by *E. coli* strain (*E. coli* cps2) without EUP pathway. In the opposite vein, it is also of interest to understand if cpsADH contributes to ethanol utilization. This facet of the problem could be probed by the use of an *E. coli* strain (*E. coli* eth2) with the EUP pathway only. Finally, *E. coli* ethcps2 with both EUP pathway and cpsADH enzyme would also be used. Both LB and M9 medium would be used to assess the relative growth performance that different *E. coli* strains could achieve in the two growth media.

Results from the series of anaerobic growth experiments revealed that *E. coli* eth2 could achieve comparable extent of growth with *E. coli* ethcps2 in both LB and M9 medium supplemented with ethanol (Figure 25a). This shows that the EUP pathway is functional and could deliver similar growth potential to the respective strains. On the other hand, *E. coli* cps2 with lacks EUP could only achieve lower levels of growth in LB and M9 medium supplemented with ethanol compared to *E. coli* eth2 and *E. coli* ethcps2. Such a result validated the theoretical postulation that endogenous ethanol utilization capability alone could not deliver high levels of growth. In general, cells grew to a higher optical density in LB medium compared to M9 medium due in large part to the presence of other catabolizable nutrients in LB medium. On the other hand, all the strains investigated achieved higher levels of growth in M9 medium supplemented with glucose compared to the same medium supplemented with ethanol. This indicates that EUP pathway in combination with endogenous ethanol utilization could not deliver comparable amount of growth as compared to endogenous glucose utilization under anaerobic conditions.

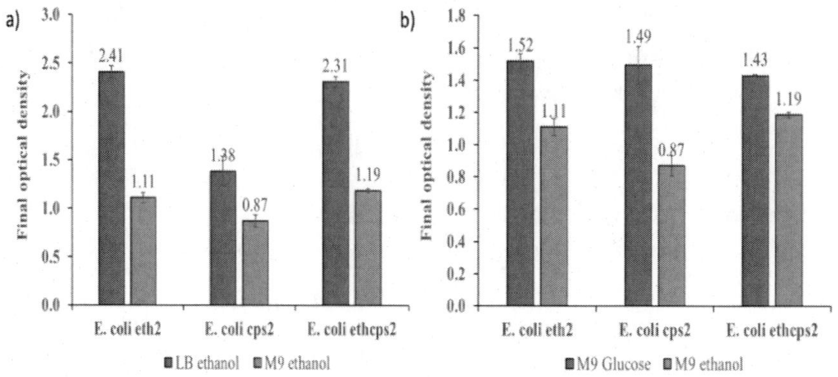

Figure 25. Anaerobic growth performance of *E. coli* eth2, *E. coli* cps2 and *E. coli* ethcps2. a) Comparison of growth performance of different strains in LB and M9 medium supplemented with ethanol, b) Assessing the relative growth potential of glucose and ethanol in M9 medium for different *E. coli* strains.

5.9. Exploring Growing Cells as an Alternative Mode for Acetophenone Biotransformation

Compared to resting cells, growing cells should provide a driving force for ethanol utilization and greater NADH regeneration given that acetyl-CoA at the end of EUP pathway is an essential metabolite siphoned off for biomass formation. Hence, growing cells should theoretically provide more NADH for pushing the biotransformation reaction towards product formation and higher yield. But, this could only work if there are no other limiting process that hamper on product formation. One such possibility is the potential limit in which acetophenone could be taken up by the cells.

Experiments were initiated to test if growing cells is a viable option for acetophenone biotransformation. This stems from the possibility in which exposure to n-hexadecane and acetophenone may be detrimental for cell growth, thereby, sowing the seeds for experimental verification of the biotransformation system. *E. coli* ethcps2 with autoinducible pthrC3 controlling EUP and cpsADH was cultivated in LB medium for growing cell biotransformation. Results revealed that growing cell is a feasible option for acetophenone biotransformation. Specifically, there was substantial production of 1-Phenylethanol in cells fed with glucose and ethanol co-substrate (Figure 26b). However, the results also indicated that a co-substrate such as glucose or ethanol is needed for biotransformation occur, which suggests that LB medium could not help generate sufficient NADH to cater to both cell growth and biotransformation requirements. Comparing between glucose and ethanol as substrates for NADH regeneration revealed that yield was higher for cells exposed to ethanol compared to glucose (Figure 26a). This suggests that the heterologous EUP pathway was more efficient than endogenous glucose metabolism in regenerating NADH for biotransformation. In direct correspondence with yield data, cells fed with ethanol also produce more 1-Phenylethanol compared to those fed with glucose (Figure 26b). Overall, the biotransformation process was relatively slow given the continued synthesis of 1-Phenylethanol into the second day of biotransformation.

To discern if diffusion of acetophenone into cells is the limiting factor for growing cell biotransformation, acetophenone concentration profiles could provide some clues into the relative importance of cellular driving force and diffusion for acetophenone uptake. Observation of similar acetophenone concentration profile across cells fed with different cosubstrates revealed that transport of acetophenone into cells is likely the limiting factor, and is not correlated with the different driving force generated by the utilization of different cosubstrates (Figure 26c). In this case, driving force generated by co-substrate utilization refers to the NADH pool synthesized that could potentially drive acetophenone conversion. However, it is important to remember that similar uptake of acetophenone does not imply similar conversion of acetophenone to 1-Phenylethanol. Finally, acetate concentration profiles also correlated with those of 1-Phenylethanol, thereby, indicating tight coupling between the biotransformation reaction and EUP pathway (Figure 26d). This then translates to the understanding that NADH pool generated by cofactor regeneration systems drives the conversion of acetophenone to 1-Phenylethanol. Similar to 1-Phenylethanol concentration profile, acetate was also produced in higher quantities in cells fed with ethanol compared to glucose, which points to proper functioning of the EUP pathway in converting ethanol to acetyl-CoA and partially on to acetate. Note that more acetate may be produced and secreted if the cells pool of acetyl-CoA was not partially depleted by growth processes.

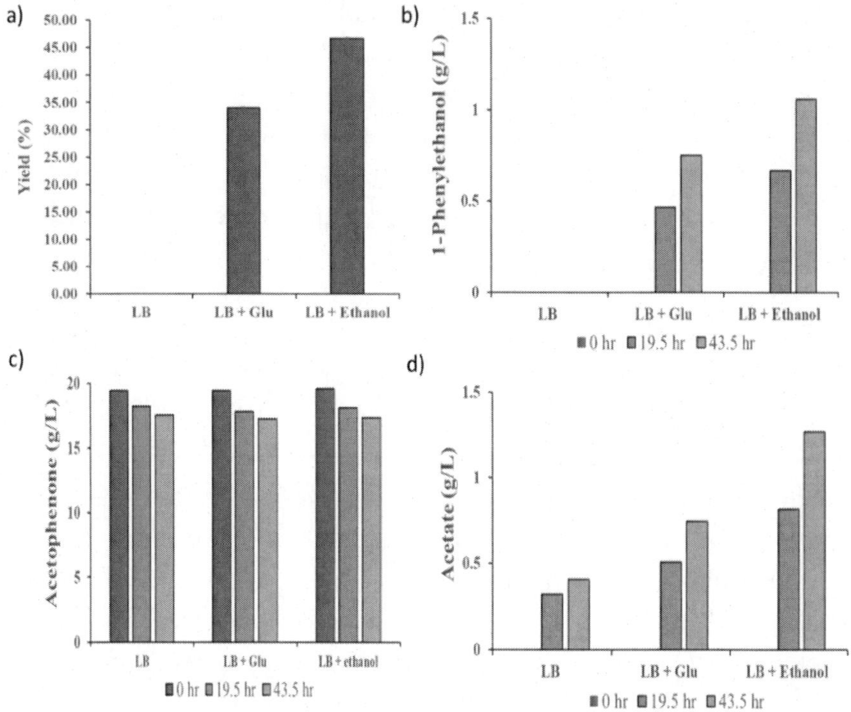

Figure 26. Growing cell mediated acetophenone biotransformation. a) Higher yield was achieved with ethanol as cofactor regeneration substrate compared to glucose in engineered cells, b) 1-Phenylethanol concentration profiles tally with those of yield, showing high product titer in cells fed with ethanol compared to glucose, c) similar acetophenone concentration profiles across categories show that co-substrate only played a small additional role in mediating uptake of acetophenone, thereby, suggesting that the driving force for acetophenone uptake may be innate transport mechanism across the *E. coli* cell envelope, d) acetate formation was highest in cells fed with ethanol compared to glucose, which is a result of the coupling between biotransformation and EUP where 1-Phenylethanol formation creates a demand for NADH that is met by conversion of ethanol to acetyl-CoA and on to acetate. However, growing cells do diverted part of the acetyl-CoA flux for biomass formation.

5.10. Resting Cells Biotransformation for Examining if Provision of Co-Substrate Could Engender Substantial Biotransformation

Experiment effort shifted from acetophenone to butanone to help establish the bioprocessing difficulties that could arise in single phase biotransformation. Specifically, water-insoluble acetophenone requires a two-phase biotransformation system that adds complexity to the process. On the other hand, a single aqueous phase biotransformation is easy and less complicated but faces challenges in substrate toxicity. To understand the efficiency in which the two gene ethanol utilization pathway (EUP) could be used to provide NADH needed for biotransformation of ketone to alcohol in a single phase system, a water soluble ketone, butanone, was selected as substrate.

Resting cells biotransformation was selected as initial experiment to assess the capability of the co-substrate to afford NADH regeneration and thus biotransformation. In essence, this set of experiments is a functionality test for the entire whole cell-based *Escherichia coli* biotransformation system. To test the effect of growth medium on biotransformation efficiency through engendering a conducive gene expression pattern for biotransformation, LB and M9 medium supplemented with either glucose or ethanol would be used for precultivation of the cells prior to biotransformation. Glucose is chosen for comparison as it is both a positive control as well as a cheap substrate that likely would elicit endogenous glucose utilization pathway to regenerate NADH in support of biotransformation reactions. Concentration chosen for both co-substrates is 10 g/L. The strain chosen for the experiments is *E. coli* ethcps2 with both EUP pathway and cpsADH.

Results from the experiments revealed that ethanol elicited higher NADH regeneration compared to glucose (Figure 27a). This indicates that the EUP pathway is more efficient in NADH regeneration compared to the endogenous glucose utilization system in *E. coli*. More importantly, the data also shows that a co-substrate such as glucose and ethanol is needed for NADH regeneration to occur given that no product formation was observed in cells suspended in phosphate buffer without co-substrate. From another perspective, the same data also suggests that the endogenous pool of NADH in resting cells is not sufficient to support butanone biotransformation without regeneration of cofactor triggered by a co-substrate such as glucose or ethanol.

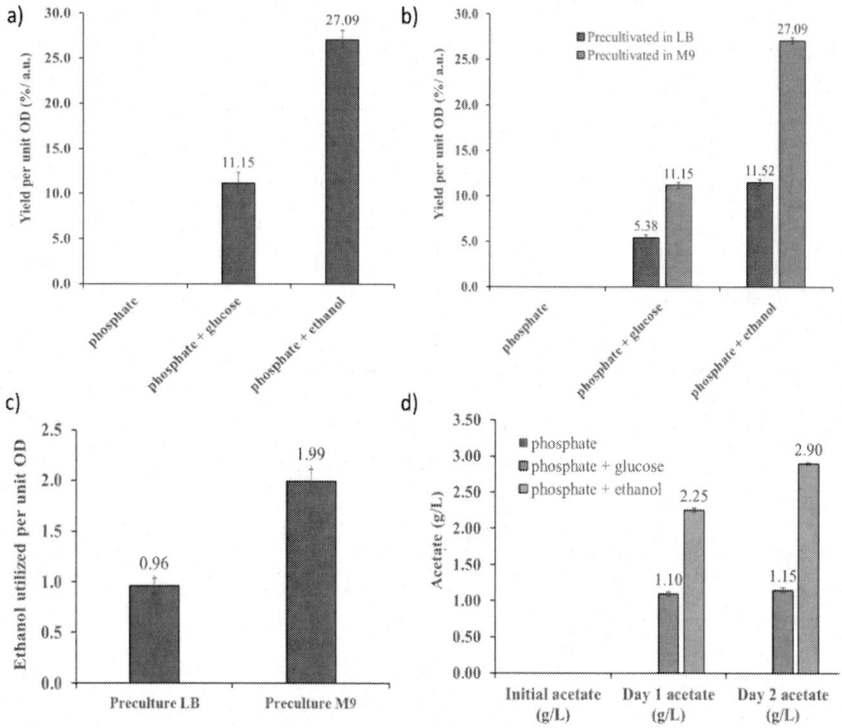

Figure 27. Resting cells anaerobic butanone biotransformation for examining if a co-substrate could power NADH regeneration and biotransformation. a) Ethanol with EUP engendered higher biotransformation efficiency for cells precultivated in M9 medium, b) Cells precultivated in M9 medium delivered higher biotransformation efficiency compared to those precultivated in LB medium., c) Preculture in M9 also results in higher ethanol utilization per unit OD probably due to a gene expression pattern conducive for ethanol uptake, d) Acetate formation was higher in cells fed with ethanol compared to glucose in resting cell biotransformation (cells precultured in M9 medium), which affirms that the EUP pathway is function with acetyl-CoA produced partially channelled to acetate in resting cells.

Another significant finding that emanates from the data is the observation that cells precultivated in M9 medium delivered higher biotransformation efficiency compared to those precultivated in LB medium (Figure 27b). The effect is present for both glucose and ethanol co-substrate, and is thus not likely to be due to the training effect of cultivating cells in M9 supplemented with ethanol prior to a biotransformation experiment. One reason that could explain the results would be the induction of a conducive gene expression pattern in *E. coli* cells during culture in M9 + 10 g/L ethanol medium that facilitates biotransformation through as yet unknown mechanisms. Besides enhanced biotransformation efficiency, preculture in M9 medium was also observed to correlate with higher ethanol utilization per unit OD compared to preculture in LB medium (Figure 27c). This result partially explains why precultivation in M9 medium would result in higher biotransformation efficiency, but it leaves open the question of why preculture in M9 medium would enhance biotransformation efficiency of cells fed with 10 g/L glucose co-substrate in phosphate buffer given that similar glucose uptake rates were observed in cells preculture in LB or M9 medium (data not shown). Finally, cells fed with ethanol engendered higher acetate production compared to those fed with glucose (Figure 27d). First, production of acetate in resting cells could be the funnelling of acetyl-CoA flux to acetate given the lack of cellular growth processes that consume acetyl-CoA, a key cellular building block. From another perspective, the result provides further confirmation that the EUP pathway is functional and well-integrated with endogenous metabolism even in resting cells in a nutrient depleted environment. Second, relative lack of acetate formation in cells fed with glucose suggests that endogenous glucose utilization system may be channelling the carbon flux from glucose efficiently to other pathways and less carbon is lost through acetate formation. Overall, this set of experiments firmly demonstrates that the EUP pathway is functional, and is able to efficiently regenerate NADH in support of butanone biotransformation in resting cells.

While encouraging, the above result must be taken in light of the observation that a fundamental limit exists in 2-butanol titer. Specifically, a maximal product titer of 4.92 g/L was obtained for resting cell butanone biotransformation for cells precultivated in either medium with ethanol as biotransformation co-substrate. This points to the existence of a tolerance limit for 2-butanol in *E. coli* ethcps2, which needs to be overcome to further improve product titer and yield per unit OD. Besides a limit for 2-butanol titer, the fundamental limit on ethanol utilization was again observed for cells precultivated in either growth medium. A similar constraint does not exist for glucose utilization most probably due to the utility of glucose in activating many associated pathways in central carbon metabolism that lead on from initial glucose catabolism. These pathways synthesize building blocks and harness energy from glucose and its downstream metabolites.

5.11. Quantitative Examination of Substrate Toxicity Effect in Single Phase Biotransformation

Cell growth tends to be more sensitive to toxicity effect of substrate; thus, resting cells were used in quantitative evaluation of the toxicity effect of butanone in biotransformation. Initial experiments aiming to examine the effect of substrate concentration on biotransformation efficiency revealed that substrate toxicity played a major role in constraining biotransformation efficiency. Specifically, 20 g/L butanone was found to inhibit biotransformation. In general, yield per unit of optical density declined with increasing substrate concentration in a monotonic fashion, which is in sync with observations of substrate toxicity effect exerted by 20 g/L butanone (Figure 28). More importantly, substrate toxicity effect was observed in resting cells precultivated in either LB or M9 + 10 g/L ethanol medium. On the other hand, product titer of 2-butanol increased with higher substrate concentration, but as the figure of merit for assessing biotransformation performance is yield, substrate toxicity problem remains important to the attainment of high product titer.

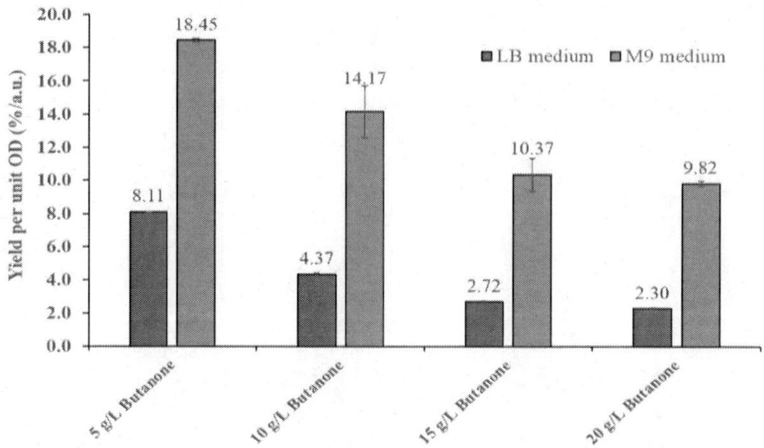

Figure 28. Effect of butanone concentration on resting cell biotransformation efficiency. Results clearly indicate existence of substrate toxicity effect in single phase resting cell biotransformation system at high butanone concentration.

Effect of growth medium on biotransformation efficiency was also assessed given the hypothesis that cells precultivated in medium with ethanol as the main carbon source would have gene expression patterns conducive for the uptake of ethanol during the biotransformation phase. In experiments comparing the performance of cells cultivated in LB and M9 medium with 10 g/L ethanol, results revealed that *E. coli* ethcps2 cultivated in M9 + 10 g/L ethanol medium delivered slightly higher product yield. This likely came about due to the induction of a gene expression pattern conducive for the uptake of ethanol.

5.12. Growing Cell Biotransformation Provides a Driving Force for Ethanol Uptake and Utilisation

Growth processes generally provide important driving forces for biochemical processes in the cell. Thus, could cell growth help improve biotransformation of butanone? Experiments were conducted with *E. coli* ethcps2 cultivated in LB and M9 + 10 g/L ethanol medium to assess the relative efficacy of the two growth media in promoting cell growth and biotransformation of butanone. Further, for each growth medium, different carbon sources such as glucose or ethanol were assessed for their efficacy in NADH regeneration that supports biotransformation.

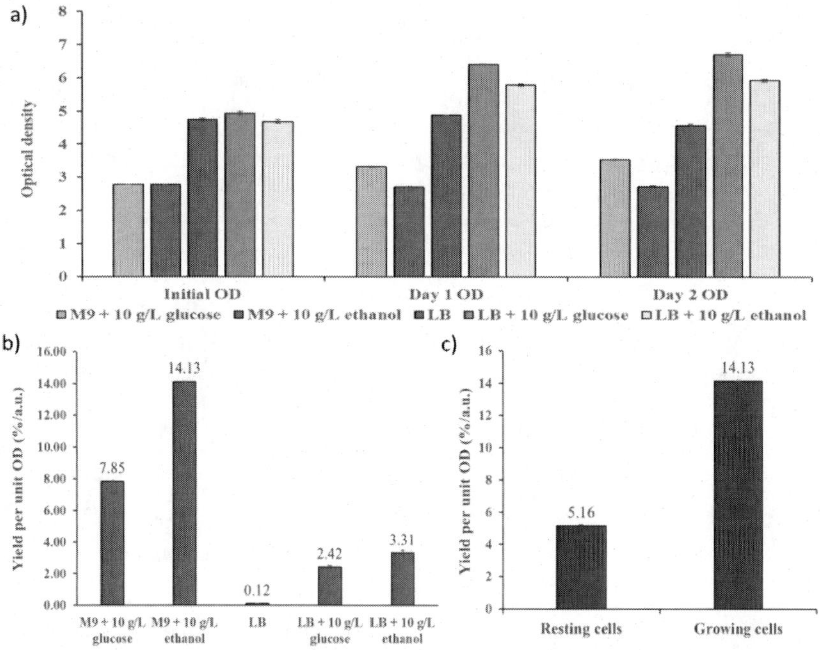

Figure 29. Examining the effect of growing cell on butanone biotransformation. a) *E. coli* ethcps2 was better able to grow with glucose as supplementary carbon source compared to ethanol, b) Cells cultivated in M9 ethanol medium outperform those cultivated in LB medium in biotransformation efficiency, c) Growing cells delivered higher yield per unit OD due largely to the development of a driving force for ethanol consumption.

Results revealed that cell growth (measured by optical density at 600 nm) was higher with glucose as supplementary carbon source in both LB and M9 medium compared to ethanol (Figure 29a). In terms of product yield per unit OD, cells grown in M9 medium had better performance compared to those cultivated in LB medium irrespective of the carbon source supplemented. This suggests that differing gene expression patterns of cells cultivated in LB and M9 medium likely played an important role in the performance of whole cell biocatalyst in butanone biotransformation. Comparison of the relative performance of different carbon sources in NADH regeneration revealed that ethanol was more effective compared to glucose based on the effect of carbon source on product yield per unit OD for both LB and M9 medium (Figure 29b). Such observation is likely a manifestation of the effect of the heterologous ethanol utilization pathway that enables a higher efficiency in cofactor regeneration compared to endogenous glucose dehydrogenase enzymes. Finally, growing cell biotransformation enhanced biotransformation efficiency compared to resting cell biotransformation (Figure 29c); thereby, reaffirming the importance of cell growth in providing the driving force for NADH generation, which is a key factor accounting for higher butanone consumption and 2butanol production.

Despite encouraging results from the use of growing cells in butanone biotransformation, lingering problems that affected resting cell biotransformation also persist in growing cell biotransformation. One of which is the apparent constraint on the amount of butanone consumed. Specifically, only a 30-50% of available butanone substrate was consumed despite the use of growing cells for biotransformation (Figure 30). This points to two possible reasons hampering butanone uptake. Firstly, toxicity of butanone on cells might have prevented greater uptake of the substrate. Secondly, native transporter in *E. coli* able to assimilate butanone might not be fully functional in substrate uptake. Such constraint on the uptake of butanone places a limit on product formation, which affects the yield of the process. Hence, further engineering is needed to help improve substrate uptake and increase product yield. One approach is through engineering a better transporter for butanone. While this is conceptually feasible, it nevertheless requires extensive work in identifying the correct transporter that uptake butanone as well as protein engineering work to improve its binding for butanone and mediating its transporter. Another approach is the creation of a driving force for butanone uptake. Such a driving force could be generated using growth to drive acetyl-CoA consumption, ethanol conversion and NADH regeneration. Large pool of NADH would subsequently drive butanone consumption and uptake.

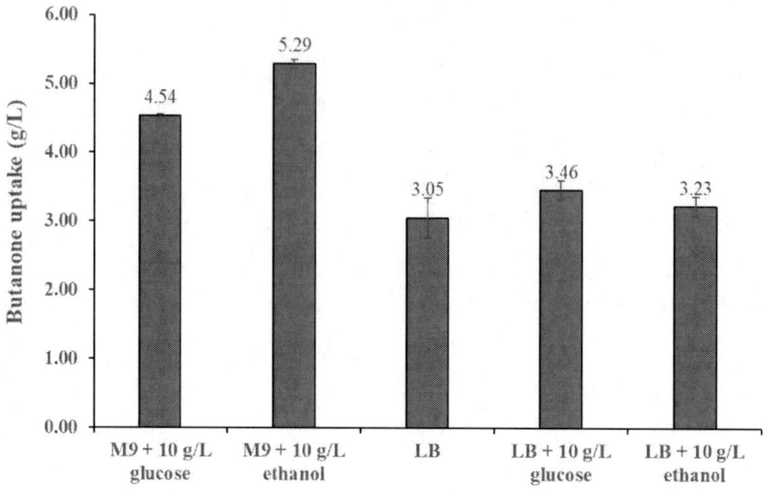

Figure 30. Only a fraction of butanone substrate was consumed for growing cell biotransformation; thereby, opening up avenue for improving uptake of butanone through transporter engineering or creating a driving force for its assimilation. Original butanone substrate concentration was about 10 g/L for this set of experiments.

Finally, similar to resting cell biotransformation, growing cell biotransformation experiment also exhibited a fundamental limit on the amount of ethanol utilized. Specifically, only about 2-3 g/L of ethanol was utilized over 2 days compared to 4-5 g/L of glucose utilized by the same strain over the same period (Figure 31a). Given that this limit on ethanol utilization also exists in the same strain undergoing anaerobic cell growth without exposure to butanone (Figure 31b), accumulation of the toxic pathway intermediate, acetaldehyde, could be an important factor hampering further ethanol utilization and placing a limit on product yield. One approach for alleviating the cytotoxic effect of acetaldehyde accumulation would be to overexpress acetaldehyde dehydrogenase to convert the pathway intermediate into acetate. On the other hand, possibility also exists of improving the expression of *ada* enzyme for converting acetaldehyde into acetyl-CoA.

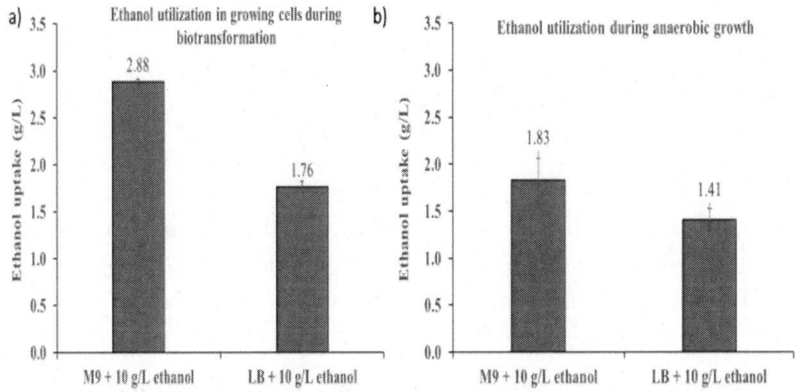

Figure 31. Ethanol utilization remains limited in growing cells, thereby, placing a fundamental limit on biotransformation efficiency. Cause of the constraint likely comes from an anaerobic biotransformation environment given that ethanol utilization was also poor in the same *E. coli* strain under anaerobic growth condition. a) Ethanol utilization in growing cells during biotranaformation, b) ethanol utilization during anaerobic growth in *E. coli* ethcps2.

5.13. Improved Growing Cell Butanone Biotransformation Experiment with Smaller Inoculum Size

One limitation of the above growing cell biotransformation experiment was that the inoculum was too large; thereby, presenting little room for the cells to grow. Without growth, there would not be a driving force to push ethanol carbon flux towards acetyl-CoA. Thus, the growing cell experiment was repeated with a smaller inoculum for both growth in LB and M9 + 10 g/L ethanol medium. Two co-substrates, glucose and ethanol, were used to assess their relative efficacy in activating NADH regeneration systems. Results revealed that growth of *E. coli* ethcps2 in M9 + 10 g/L ethanol medium generated a higher yield per unit OD (48%) compared to growth in LB medium supplemented with ethanol (Figure 32a). More importantly, ethanol fared better as a co-substrate for NADH regeneration compared to glucose given the higher yield per unit OD generated. This comes about likely due to the dual NADH regeneration systems available in the endogenous *adhE* system for ethanol utilization as well as the heterologous ethanol utilization pathway. But, at a broader level, M9 medium supplemented with ethanol may be better at activating NADH regeneration pathways endogenous to the cell; and thereby, providing an enhanced platform for cofactor regeneration leading to higher yield per unit OD.

While the initial inoculum was able to expand into a large cell population in the redesigned growing cell biotransformation experiment, constraints observed in the earlier incarnation of the same experiment remains. One example is the fundamental limit on the amount of ethanol utilized (Figure 32b), which has been attributed to the toxicity arising from acetaldehyde accumulation due to poor kinetic properties of aldehyde dehydrogenase or poor expression of the detoxification enzyme. Another issue concerns the diversion of ethanol carbon flux to acetate through the acetaldehyde to acetate shunt. While production of acetate through aldehyde dehydrogenase also generates one NADH per turn of reaction, the relative utility of this reaction compared to driving all ethanol carbon flux to acetyl-CoA in support of cell growth remains uncertain. Overall, a detailed investigation is needed before a firm conclusion could be arrived for the relative merits of channeling ethanol carbon flux to acetate or acetyl-CoA in supporting butanone biotransformation.

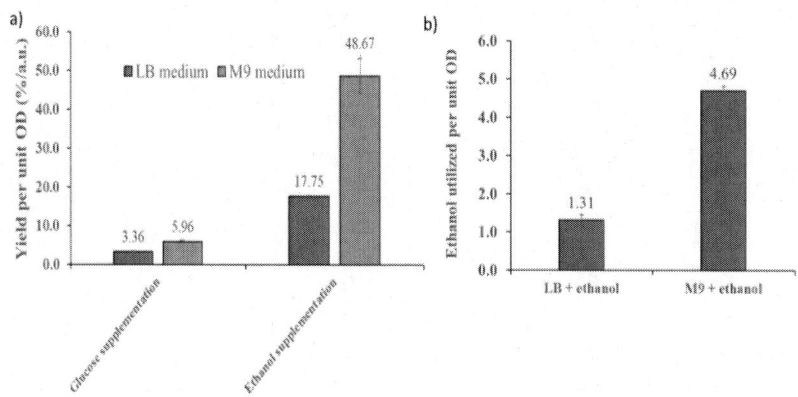

Figure 32. Reduced inoculum growing cell biotransformation for *E. coli* ethcps2. a) Ethanol provides a higher biotransformation efficiency together with a growth medium effect that favours M9 over LB medium, b) Fundamental limit in ethanol utilization remains despite growing cell inducing a driving force for ethanol consumption.

Another lesson that we can learn from the ethanol utilized per unit OD data is that M9 medium definitely offers a gene expression pattern conducive for the uptake and utilization of ethanol compared to LB medium supplemented with ethanol (Figure 32b). This comes from observation of ethanol utilized per unit OD of 4.69 in cells grown in M9 ethanol medium compared to 1.31 in cells grown in LB ethanol medium. Such a finding indicates that M9 medium may be the preferable medium for EUP-based NADH powered butanone biotransformation.

The primary motivation for exploring growing cell butanone biotransformation was to examine the hypothesis if cell growth could provide a driving force that help improve NADH regeneration and biotransformation efficiency through increased ethanol utilization. Results illustrated in Figure 33 shows that growing cells do provide a higher biotransformation efficiency compared to resting cells (Figure 33a). More importantly, the higher biotransformation efficiency achieved in growing cells could arise from enhanced ethanol utilization per unit OD from growing cells compared to resting cells (Figure 33b). Thus, the presented experimental data fits the story of cell growth generating a driving force for ethanol utilization and NADH regeneration through continued consumption of acetyl-CoA. Enhanced NADH regeneration in growing cells subsequently helped improved biotransformation efficiency.

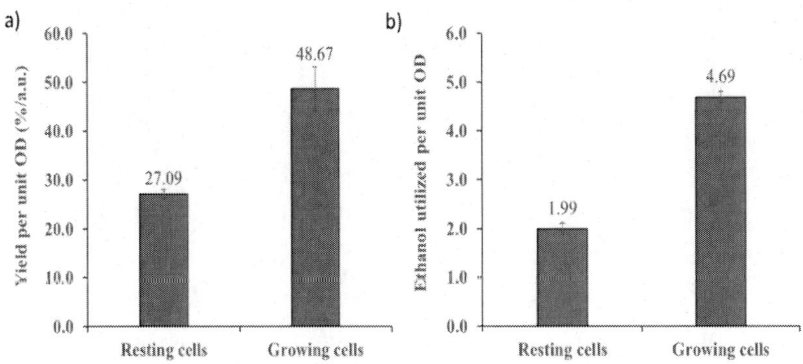

Figure 33. Growing cell biotransformation provides a driving force that results in higher biotransformation efficiency and ethanol utilization per unit OD. Cells in this comparison were either precultivated (resting cells) or cultivated (growing cells) in M9 medium.

5.14. Testing the Relative Efficacy of Endogenous Enzymes in NADH Regeneration with Either Glucose or Ethanol as Co-Substrates in Biotransformation

While it has been demonstrated that the heterologous EUP pathway works in NADH regeneration, one intriguing question concerns how efficient the endogenous glucose dehydrogenase and alcohol dehydrogenase (adhE) is in generating NADH needed to power the cpsADH catalysed conversion of butanone to 2-butanol. To this end, an *E. coli* strain with cpsADH enzyme but without ethanol utilization pathway (i.e., *E. coli* cps2) was used to test the relative efficacy of glucose and ethanol as carbon sources for supporting butanone biotransformation. Resting cell experiment was chosen to investigate the relative efficacy of different carbon sources in powering biotransformation. Experimental results revealed that the endogenous pool of NADH in *E. coli* cps2 cells was insufficient for powering butanone biotransformation with no formation of 2-butanol observed in phosphate buffer without cosubstrates. This suggests that a co-substrate such as glucose or ethanol is needed to help generate NADH for biotransformation. In cases where a co-substrate is provided, endogenous ethanol utilization provided comparable biotransformation efficiency compared to endogenous glucose utilization (Figure 34a). This suggests that the NADH regeneration capacity of endogenous ethanol utilization pathway in resting *E. coli* cps2 cells is comparable to those of the endogenous glucose utilization pathway.

Acetate formation can serve as a diagnostic tool for how well endogenous ethanol utilization pathway is functioning. In this regard, lower acetate formation in cells fed with ethanol compared to glucose suggests that endogenous ethanol utilization pathway may not be efficient in converting ethanol to acetyl-CoA (Figure 34b). Evidence supporting this postulation could be found in the poor ethanol utilization profile depicted in Figure 34c. Such a profile suggests that the half-life of endogenous adhE is relatively short and it requires renewed transcription and translation in resting cells to provide a pool of adhE enzymes for mediating ethanol conversion and NADH regeneration. This explains why there was almost no consumption of ethanol in the first day of biotransformation. Finally, glucose was well utilized over 2 day of biotransformation, but it is an inefficient mode for NADH regeneration given that it could only deliver a comparable amount of biotransformation efficiency relative to ethanol utilization (Figure 34d).

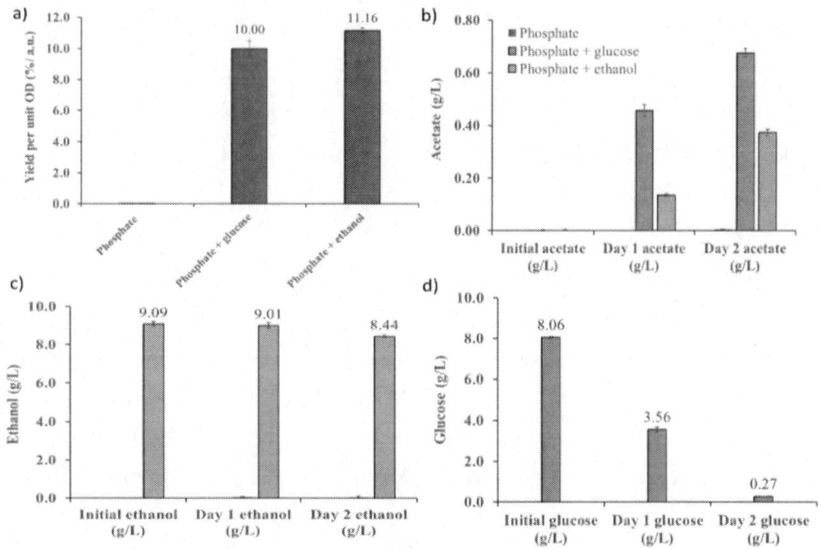

Figure 34. Endogenous ethanol utilization delivers comparable biotransformation efficiency compared to endogenous glucose utilization in resting *E. coli* with cpsADH but without EUP pathway. a) Yield per unit OD for cells precultivated in M9 medium but fed either glucose or ethanol as co-substrate for biotransformation, b) acetate formation was lower in cells fed with ethanol compared to glucose, which suggests endogenous ethanol utilization may not be efficient in ethanol conversion, c) relatively poor ethanol utilization dovetails with lack of driving force for ethanol utilization, d) glucose was well utilized by resting *E. coli* cps2 cells.

Given that cell growth has been shown to provide a driving force for pushing ethanol carbon flux towards acetyl-CoA, which, in the process, generates more NADH for supporting butanone biotransformation, *E. coli* cps2 without ethanol utilization pathway was further used in growing cell biotransformation experiment to assess the relative efficacy of glucose and ethanol in NADH regeneration. In experiments conducted with LB medium as base medium, *E. coli* cps2 fed with ethanol was more efficacious in converting butanone to 2-butanol; thereby, providing a higher yield per unit OD (Figure 35a). This indicates that endogenous adhE could utilize ethanol for NADH regeneration, and more importantly, endogenous ethanol utilization was more effective in cofactor regeneration compared to glucose utilization. What is interesting from the experiments is the observation of a fundamental limit on ethanol utilization similar to that present in *E. coli* with ethanol utilization pathway (Figure 35b). Given that both endogenous adhE and ethanol utilization pathway have identical pathway architecture for ethanol catabolism, it is plausible that similar mechanisms are at play in limiting ethanol utilization. One leading hypothesis is the accumulation of acetaldehyde pool as described earlier. Limits on ethanol utilization imposed by anaerobic environment used in the biotransformation could be another reason. Acetate production was also deficient in *E. coli* cps2 fed with ethanol in LB medium compared to those fed with glucose (Figure 35c). This may arise due to the channelling of acetyl-CoA to synthesis of cellular building blocks during cell growth in *E. coli* cps2 fed with ethanol, which reduced the precursors for the formation of acetate. Another limitation observed in the experiments is the poor growth of *E. coli* cps2 in LB supplemented with ethanol compared to those cultivated in unsupplemented LB or LB supplemented with glucose (Figure 35d). This suggests the toxicity of ethanol on cell growth, and points to areas for further engineering to improve *E. coli* tolerance to ethanol.

The same growing cell experiment was repeated with M9 medium to assess for biotransformation efficiency under a more basal medium condition compared to the previously described experiment in LB medium. Results revealed that *E. coli* cps2 fed with ethanol under growing cell mode could provide marginally higher biotransformation efficiency compared to cells fed with glucose (Figure 36a). This suggests that the endogenous adhE system for ethanol utilization could deliver appreciable amount of NADH for supporting biotransformation. In addition, the result also shows that the endogenous adhE system is more efficient than endogenous

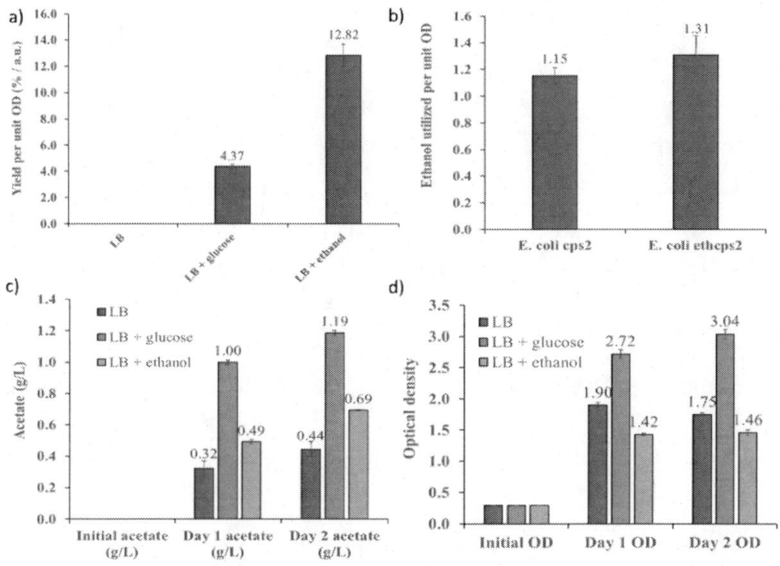

Figure 35. Assessing the endogenous NADH regeneration capability in *E. coli* cps2 during growing cell biotransformation in LB medium supplemented with different co-substrates. a) Ethanol with driving force from cell growth engenders higher biotransformation efficiency compared to glucose, b) comparable ethanol utilization was observed in *E. coli* cps2 and *E. coli* ethcps2, c) Cells fed with ethanol produce less acetate compared to those fed with glucose, which dovetails with channelling of acetyl-CoA to cell growth in *E. coli* cps2 fed with ethanol, d) Cells fed with ethanol in LB medium achieved lower optical density than those fed with glucose, and suffers from cell retardation due to ethanol toxicity.

glucose utilization in regenerating NADH largely due to the diversion of metabolic flux from glucose to associated pathways that mediates its catabolism. One surprise from the experiment is the lower than expected growth of *E. coli* cps2 in M9 ethanol medium compared to the same strain in the same medium without infusion of butanone (Figure 36b). Given that the same strain could grow to appreciable optical density when fed with glucose and butanone, toxicity of butanone should not be a reason for the observed poor growth in M9 ethanol medium. Rather, inability of the endogenous adhE enzyme in detoxifying acetaldehyde under the influence of butanone may be the ultimate cause explaining the poor growth of *E. coli* cps2 in M9 ethanol medium. Quantification of ethanol utilization shows that the *E. coli* cps2 strain without EUP pathway consumes about

1.5 g/L of ethanol over 2 days compared to the 2-3 g/L of ethanol consumed by *E. coli* ethcps2 with EUP pathway over 2 days (Figure 36c). This shows that the ethanol utilization capacity of the endogenous adhE enzyme is not particularly high. Finally, no acetate was produced in *E. coli* cps2 fed with ethanol in what is a major surprise of the experiment. Such a phenomenon could arise as the cells utilize most of the acetyl-CoA derived from ethanol catabolism to support cell growth rather than channelling it for acetate formation. In essence, the *E. coli* cps2 cells in this experiment system did not have a large intracellular pool of acetyl-CoA and suggests that the cells is in an energy deficient state.

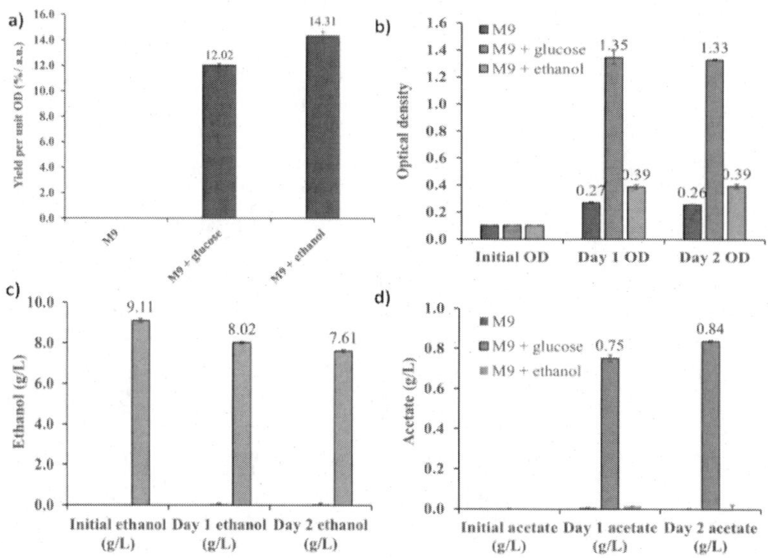

Figure 36. Assessing endogenous NADH regeneration capability in *E. coli* cps2 cultivated in M9 medium under growing cell biotransformation. a) Endogenous ethanol utilization help deliver marginal improvement in biotransformation efficiency compared to endogenous glucose utilization, b) cells fed with ethanol could hardly grow most probably due to inability to detoxify acetaldehyde intermediate under biotransformation conditions, c) Limit exists in the utilization of ethanol by *E. coli* cps2, suggesting that endogenous ethanol utilization pathway could not consume large amount of ethanol under anaerobic conditions, d) Acetate was not produced by *E. coli* cps2 under ethanol feeding conditions most probably due to the diversion of most acetyl-CoA to kickstart cell growth in lieu to channelling to acetate formation.

5.15. Effect of Ethanol Concentration on Product Yield in Butanone Biotransformation

Ethanol is not a native substrate of *E. coli*, even though the bacterium possesses an alcohol dehydrogenase, adhE, that could metabolize ethanol to acetyl-CoA. More importantly, many studies have investigated the ethanol tolerance of *E. coli* and have shown that ethanol exerts a toxicity effect on the species [38-41]. Hence, efforts were also directed to investigate whether a lower ethanol concentration could facilitate a higher product yield to be obtained, which is one of the yardsticks for assessing whether ethanol induced cellular toxicity that resulted in lower biotransformation yield. Results revealed that higher ethanol concentration enabled a greater product yield to be obtained and a direct correlation could be obtained between ethanol concentration and product yield per unit OD (Figure 37). The observed effect is independent of the growth medium used for precultivating the resting cells prior to biotransformation.

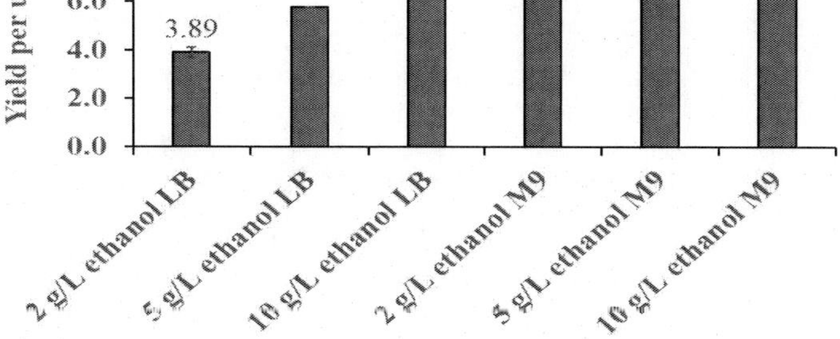

Figure 37. Effect of ethanol concentration on biotransformation efficiency in resting *E. coli* ethcps2 precultivated in either LB or M9 + 10 g/L ethanol medium.

This dovetails with theoretical predictions from assessing the EUP pathway given that a higher ethanol concentration would push more carbon flux down the pathway and enable the generation of higher NADH concentration. More important, there are signs of the growth medium effect where precultivation in M9 + 10 g/L ethanol medium elicited a higher biotransformation efficiency compared to preculture in LB medium. This could be due to "training' of *E. coli* ethcps2 to use ethanol via eliciting a gene expression pattern conducive for the import and catabolism of ethanol.

5.16. Analysis of Carbon Flux through the Ethanol Utilization Pathway

Substrate utilization is an important metric for any biotransformation system. In this area, ethanol as co-substrate enabled efficient utilization of butanone through the conversion of the substrate into product, thereby, presenting a driving force for substrate utilization. Similarly, acetate production is another important metric for assessing the carbon flux flowing from ethanol. Specifically, the ethanol utilization pathway bifurcates at acetaldehyde to either acetate or acetyl-CoA. HPLC analysis revealed substantial acetate production. On a molar basis, half of ethanol utilized was converted to acetate, with the other half presumably flowing to acetyl-CoA (Figure 38). This result provides an interesting contrast to acetate production detected during growth of *E. coli* ethcps2 in M9 + 10 g/L ethanol medium. Specifically, acetate production was substantially lower during growth of *E. coli* ethcps2 on ethanol. To this end, growing cell biotransformation experiments may be able to provide further insights into how cell regulatory mechanisms govern the production of acetate during butanone biotransformation. Overall, given that aldehyde dehydrogenase that converts acetaldehyde to acetate also produces one NADH per turn of reaction, total NADH production through ethanol utilization should be similar despite the diversion of ethanol carbon flux to acetate.

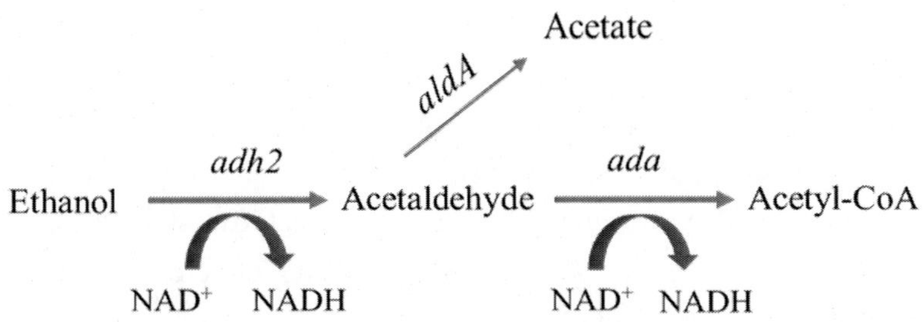

Figure 38. As a pathway intermediate, acetaldehyde is also the bifurcation point for ethanol carbon flux. Estimates suggest that on a molar basis, approximately half of the ethanol flux flows to acetate, while the other half flows to acetyl-CoA.

5.17. Using Adaptive Laboratory Evolution for Generating an *E. Coli* Strain Better Able to Utilize Ethanol

Given poor ethanol utilization during biotransformation, efforts were directed towards developing a new strain of *E. coli* ethcps2 harbouring both ethanol utilization pathway and cpsADH enzyme with better ethanol utilization properties. To this end, a long-term adaptive evolution experiment was started in which *E. coli* ethcps2 was cultivated under ethanol selection pressure to allow adaptive evolution to select for mutants able to utilize ethanol more efficiently (Figure 39). Specifically, *E. coli* ethcps2 was cultivated in M9 + 10 g/L ethanol medium. Passages were made every four days and growth performance and ethanol utilization were assessed for each even numbered passage. At passage 4, a strain of *E. coli* ethcps2 was obtained in which ethanol utilization was faster (Figure 40). However, growth of the strain was retarded, and critically, acetate production was abolished. This suggests that a new variant of *adh2* enzyme in the ethanol utilization pathway able to utilize ethanol more efficiently has emerged. In addition, the cell's ability to detoxify the acetaldehyde generated from ethanol was reduced with the emergence of variants of aldehyde dehydrogenase not able to convert acetaldehyde to acetate. Together, these new enzyme variants resulted in a cellular phenotype exhibiting enhanced ethanol utilization but with retarded growth due to the toxicity effect from accumulated acetaldehyde. This observation also provides a different perspective of how evolution select for particular mutants. Specifically, in this case, detoxification of acetaldehyde through acetate production was abolished and this was selected to help channel flux of the ethanol utilization pathway to acetyl-CoA, but resulting accumulation of toxic acetaldehyde resulted in growth retardation. This observation is puzzling as slower growth was selected to help provide for a higher maximal growth potential.

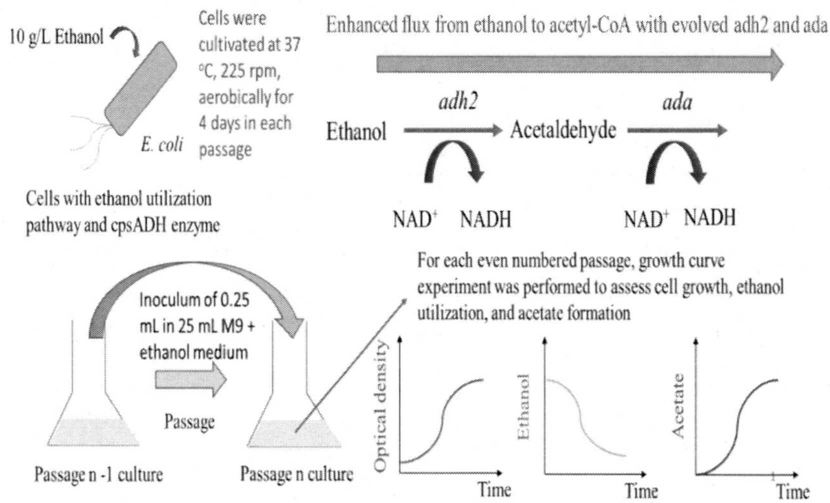

Figure 39. Schematic diagram showing the conceptual and workflow of the long-term aerobic evolution experiment to evolving higher and faster ethanol utilization capability in *E. coli* ethcps2 with EUP and cpsADH. Adaptive evolution experiment proceeds with subculture every 4 days in M9 + 10 g/L ethanol medium at 37°C and 225 rpm shaking in shake flasks. Strain from each even number passage was characterized for their growth performance, ethanol utilization capability and acetate production.

Assay for growth and ethanol utilization of the subsequent even numbered passage (passage 6) reveals a strain with slightly slower ethanol utilization and growth rate (Figure 40).

But, importantly, the strain remained unable to produce acetate under aerobic conditions (Figure 40c), which suggests that the metabolic path linking acetaldehyde to acetate remain blocked, and the only way for the cell to detoxify acetaldehyde is through the synthesis of acetyl-CoA.

Emergence of an evolved strain with slower ethanol utilization suggests a method for reducing acetaldehyde production to help reduce the toxicity effect arising from accumulating this intracellular toxic aldehyde. One possibility is a reduction in the transport of ethanol into the cell. Another possibility could be a further mutation in *adh2* leading to an enzyme with poorer ethanol conversion efficiency.

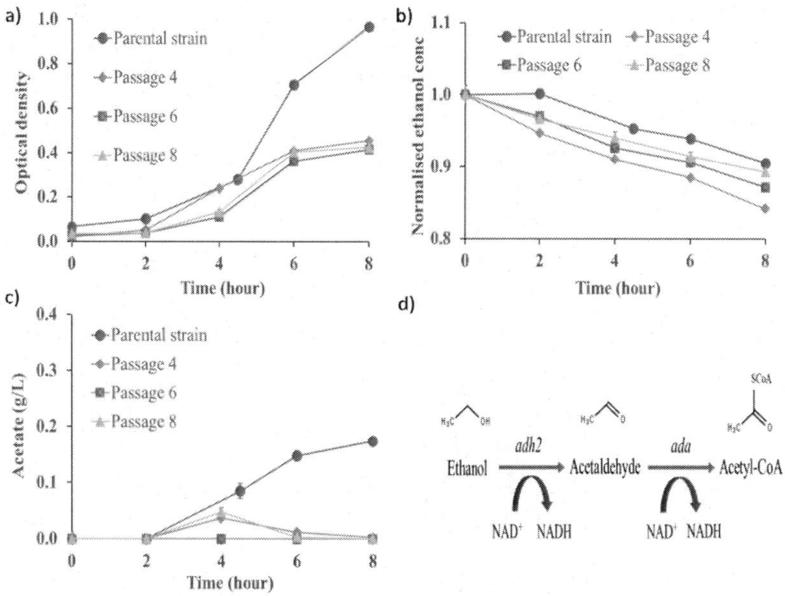

Figure 40. Growth and ethanol utilization performance of different strains obtained from aerobic long-term evolution experiment. a) Evolved strains delivered poorer growth performance, b) ethanol utilization performance was enhanced in Passage 4, but the trait reverted back to parental level in Passage 6 and 8, c) acetate production capability was significantly reduced in the evolved strains, which may be testament of reduced acetaldehyde detoxification, d) pathway architecture of the ethanol utilization pathway that is subjected to evolution.

In passage 8 of the aerobic long-term evolution experiment, strain obtained delivered similar poor growth performance in M9 + 10 g/L ethanol medium as the Passage 6 strain. While acetate production remained blocked, ethanol utilization efficiency was observed to revert back to that of the parental strain prior to the evolution experiment. Such data indicates that deleterious mutations have been induced in the genome or plasmid carrying the ethanol utilization pathway with the overriding selection pressure being the improvement in capability to detoxify accumulated acetaldehyde. Mutations could arise from two mutagenic sources: ethanol and acetaldehyde. Specifically, ethanol is known to induce multiple mutations in pathways related to membrane stress response and cellular physiology [42-44]. On the other hand, acetaldehyde is a strong oxidising agent and may cause DNA damage, that when inefficiently repaired, would result in mutations. Overall, trodding down the evolution fitness landscape towards poorer ethanol utilization and poor growth rate meant that, at passage 8, the evolved strain no longer represents a suitable chassis for further evolution. In addition, possibility also exists of the mutant being trapped in a low fitness region of a rugged fitness landscape defined by instances of high and low fitness in close proximity to each other where the troughs are evolutionary traps difficult for a strain to "escape" from. Thus, the long-term evolution experiment was terminated at this stage.

5.18. Examining the Utility of Passage 4 Mutant of *E. Coli* Ethcps2 in Resting Cell and Growing Cell Biotransformation

The strain chosen for further study comes from Passage 4 of the aerobic long-term evolution experiment. Choice of this strain for further study arise from the observation of abolishment of acetate production capability in the study presumably due to mutations in the enzyme catalysing conversion of acetaldehyde to acetate. This metabolic characteristic provides an opportunity to test the hypothesis that significant flux flowing from acetaldehyde to acetate during ethanol utilization actually provide a conduit for alleviating the toxicity of acetaldehyde accumulation in engineered *E. coli*, which ultimately results in better biotransformation efficiency. In these tests, both resting cell and growing cell biotransformations under anaerobic conditions were used to provide two perspectives of the problem. Specifically, resting cell biotransformation provides information on how the evolved EUP pathway channel ethanol flux to acetyl-CoA under the condition of no driving force from cell growth. Such insights and knowledge would help in comparing the results from growing cell experiment to assess how the evolved EUP pathway would allow ethanol flux to flow to acetyl-CoA and whether any blockage in the evolved pathway could be identified.

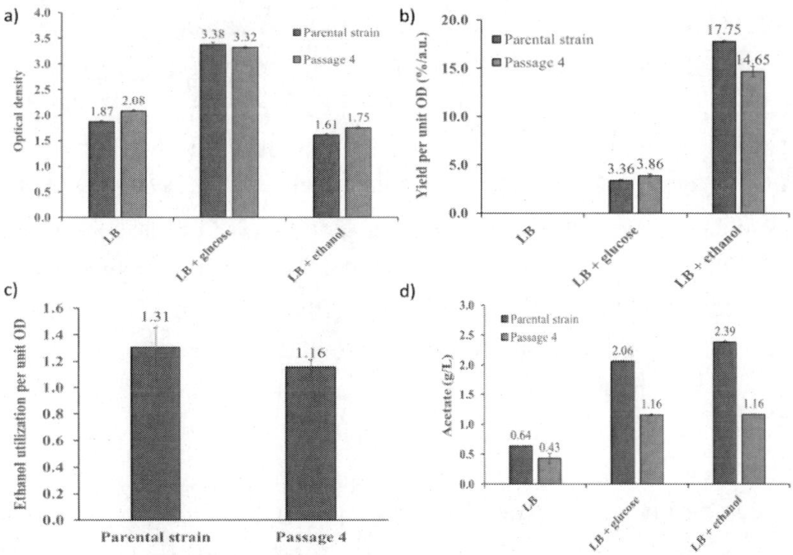

Figure 41. Growing cell biotransformation performance of *E. coli* ethcps2 Passage 4 in LB medium. a) Similar growth performance was observed in Passage 4 and parental strain in LB medium supplemented with different carbon sources, b) Passage 4 strain delivered lower biotransformation efficiency compared to parental strain, c) similar ethanol utilization was observed in Passage 4 and parental strain in LB supplemented medium, d) acetate formation was noticeably lower in Passage 4 strain compared to parental strain for both glucose and ethanol supplementary carbon source; thereby, pointing to a deficiency in acetate formation enzymes in the Passage 4 evolved strain.

In growing cell biotransformation where *E. coli* ethcps2 Passage 4 strain was fed with different carbon sources, 10 g/L butanone, and LB base medium, the evolved strain achieved similar growth performance compared to the parental strain across different supplementary substrates such as glucose and ethanol (Figure 41a). This observation could be explained by the cell's utilization of peptides and amino acids for growth, which mask the effect of an evolved EUP on cell growth. Such a deficiency could be corrected if ethanol is the main carbon source in a basal medium such as M9, which was explored in the following experiment. In terms of yield per unit OD, the evolved strain was less efficient compared to the parental strain in butanone biotransformation resulting in lower product titer (Figure 41b). In many ways, ethanol utilization performance metric would provide a useful check for the results obtained in aerobic growth experiments. But, cultivation in LB as base medium may negate cellular need for growth on ethanol, which may be the main reason why similar ethanol utilization per unit OD values were observed for Passage 4 and parental strain (Figure 41c). Acetate concentration readings from the experiment confirmed reduced acetate production in Passage 4 compared to the parental strain (Figure 41d); thereby, strongly suggesting lower acetaldehyde dehydrogenase activity in the cells. In addition, reduced acetate formation was also observed in the Passage 4 strain fed with glucose, which suggests that there may be other acetate production enzymes impacted by deleterious mutations induced by ethanol during adaptive evolution.

Comparing the biotransformation performance of *E. coli* ethcps2 Passage 4 during growing cell biotransformation in LB and M9 medium revealed that cells in M9 medium continued to outperform those in LB medium in terms of yield per unit OD (Figure 42a). This is similar to results obtained during growing cell biotransformation with the parental strain. However, yield per unit OD and optical density obtained with the evolved strain was lower compared to the parental strain under the same experimental conditions (Figure 42b and 42c). This is unexpected considering that the evolved strain has less carbon loss through the acetaldehyde to acetate shunt, and theoretically, more ethanol carbon flux was driven through to acetyl-CoA, a key intermediate for synthesizing building blocks for the cell. One reason for reduced yield per unit OD could be mutations in cpsADH enzyme leading to poorer kinetic performance. Another possibility concerns the cytotoxicity of accumulated acetaldehyde in the cell which negatively impacts on cell growth and driving force for ethanol utilization and NADH generation. Finally, analysis of ethanol utilization per unit OD reveals the surprising finding that Passage 4 strain was less able to utilize ethanol compared to the parental strain (Figure 42d). This contradicts earlier finding of enhanced ethanol utilization in Passage 4 strain over parental strain during aerobic cultivation in M9 ethanol medium. Possible reason could be the toxicity effect of butanone as well as less efficient ethanol utilization under anaerobic cultivation condition. Overall, constraints that hampered achievement of higher yield per unit OD and 2-butanol titer remains with the evolved strain and this includes: (i) poor ethanol utilization, and (ii) low butanone conversion. Finally, detoxification of acetaldehyde appears to be crucial for cellular health and is implicated in achievement of higher yield per unit OD.

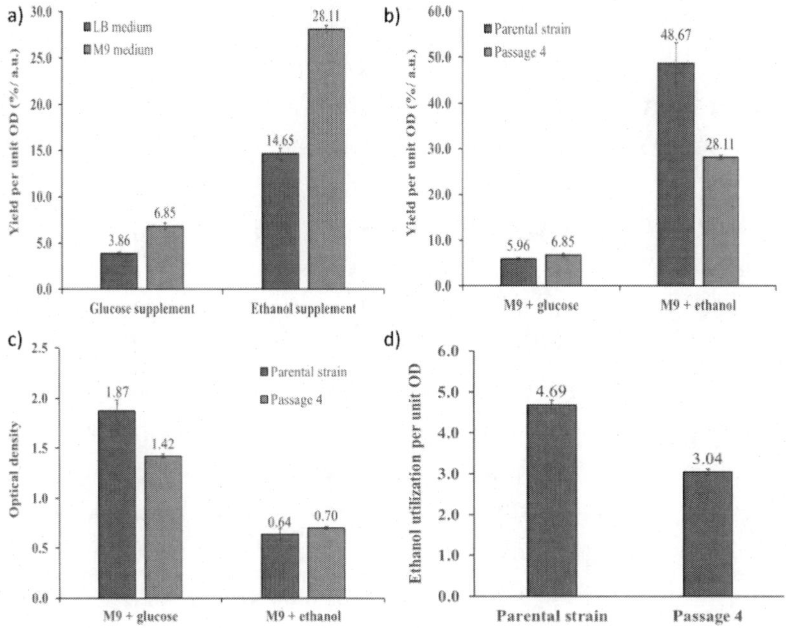

Figure 42. *E. coli* ethcps2 Passage 4 performed poorer than the parental strain in growing cell biotransformation in M9 medium. a) Similar to the parental strain, Passage 4 strain cultivated in M9 medium delivered higher biotransformation efficiency compared to LB medium for both glucose and ethanol, b) Passage 4 strain delivered lower biotransformation efficiency compared to parental strain in M9 ethanol medium, c) Lower optical density was observed for Passage 4 strain compared to parental strain in growing cell biotransformation in M9 medium, d) Ethanol utilization in Passage 4 strain was surprisingly lower than that achieved in parental strain.

To understand how the evolved EUP pathway in Passage 4 evolved strain is able to utilize ethanol without a growth-related driving force, resting cell biotransformation experiments were conducted. An ancillary goal of the experiment was in determining whether the evolved strain is more efficient in converting butanone into 2-butanol compared to the parental strain. Similar to growing cell biotransformation, the relative efficacy of glucose or ethanol in enabling NADH regeneration and biotransformation was also assessed. Results revealed that cells of Passage 4 evolved strain precultivated in LB medium delivered poorer biotransformation efficiency compared to the parental strain with ethanol as co-substrate (Figure 43a). Comparison of the co-substrates revealed that cells fed with ethanol achieved higher biotransformation efficiency compared to those fed with glucose, which reaffirms the utility of the ethanol utilization pathway in regenerating NADH (Figure 43b). However, fundamental limit in ethanol utilization remains in the evolved strain, which is a barrier to higher biotransformation efficiency. More importantly, ethanol utilization capability was observed to be lower in the Passage 4 evolved strain compared to parental strain, which is contrary to observations made in aerobic growth characterization assays of the evolved strain in M9 ethanol medium (Figure 43c). Possible reason for the discrepancy could be the anaerobic nature of this biotransformation experiment. Finally, diminished acetate production confirms earlier observations that the evolved strain had deficient aldehyde dehydrogenase unable to efficiently convert acetaldehyde to acetate (Figure 43d).

In resting cell biotransformation experiment conducted with evolved strain precultivated in M9 medium, biotransformation efficiency of the evolved strain was significantly lower than the parental strain in cases where the cells were fed with ethanol (Figure 44a). On the other hand, when glucose was used as co-substrate, biotransformation efficiency of the evolved strain was slightly higher than the parental strain. Similar to earlier experiments with resting cells, ethanol utilization remains limited (data not shown). Ultimately, relative lack of biotransformation efficiency in the evolved strain when ethanol was used as co-substrate largely comes about due to deficiency in regenerating NADH through the acetaldehyde-acetate shunt where the participating aldehyde dehydrogenase was likely mutated into a variant with poorer kinetic properties during adaptive evolution experiments. This helps explain lower acetate formation in the evolved strain relative to parental strain (Figure 44b), and the results speaks of the need for engineering a faster and more efficient route through acetaldehyde to acetyl-CoA in enabling more efficient biotransformation.

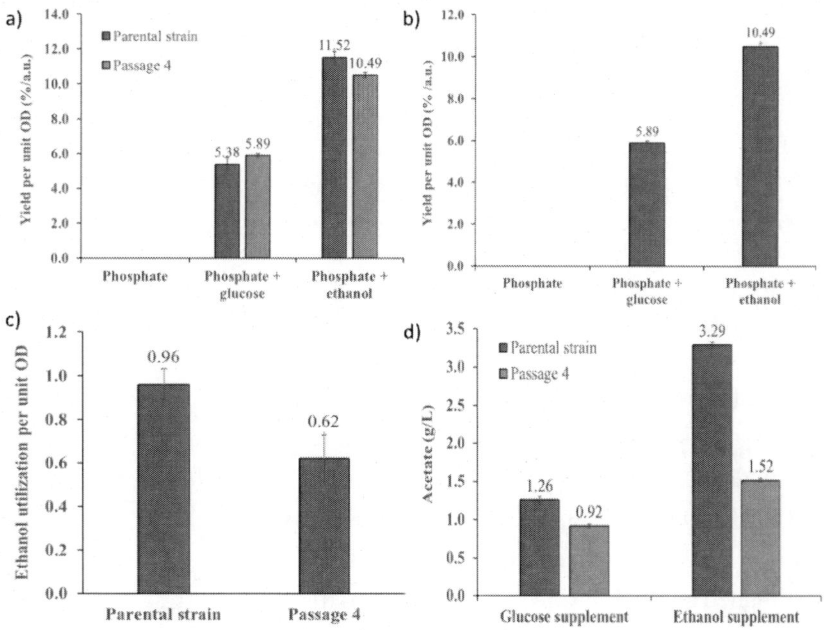

Figure 43. *E. coli* ethcps2 Passage 4 strain in resting cell mode delivers poorer biotransformation efficiency and ethanol utilization compared to parental strain under the same biotransformation conditions. a) Passage 4 evolved strain delivered poorer yield per unit OD compared to parental strain for cells precultivated in LB and supplied with ethanol during biotransformation, b) Similar to observations in parental strain, Passage 4 evolved strain fed with ethanol also delivered higher biotransformation efficiency compared to cells fed with glucose, c) Ethanol utilization per unit OD was observed to be lower for Passage 4 strain compared to parental strain, d) Acetate formation was also lower in Passage 4 strain compared to parental strain, which highlights a deficiency in detoxifying acetaldehyde during ethanol utilization.

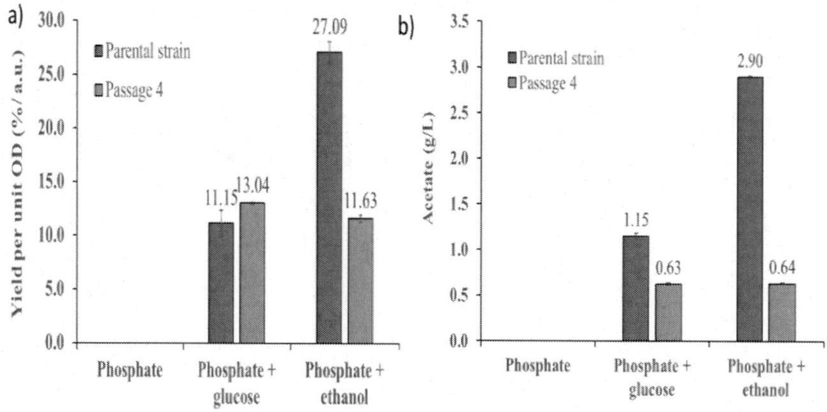

Figure 44. *E. coli* ethcps2 Passage 4 precultivated in M9 ethanol medium delivered lower biotransformation efficiency and acetate production compared to parental strain in resting cell biotransformation. a) Yield per unit OD was lower for Passage 4 cells compared to parental strain, b) Acetate formation was significantly lower in Passage 4 strain than parental strain during resting cell biotransformation, thereby, suggesting that acetaldehyde-acetate shunt may not be functioning optimally.

In summary, adaptive evolution experiment did not generate a strain with enhanced ethanol utilization properties in anaerobic biotransformation experiments. Whether resting cell or growing cell, biotransformation efficiency of the best evolved strain (Passage 4) was inferior to that achieved by the parental strain. This is correlated with diminished acetate formation capability in the evolved strain, and suggests that a deficiency exists in the acetaldehyde-acetate shunt catalysed by a native aldehyde dehydrogenase (aldA and aldB) in *E. coli*. Likely adaptive evolution experiment in a rugged fitness landscape has generated mutations in one of the endogenous aldehyde dehydrogenase in *E. coli* resulting in a deficiency in acetate formation. Lack of detoxification of acetaldehyde in resting cells may have limited NADH regeneration and thus limited biotransformation efficiency. Given the toxicity of acetaldehyde, growing cells of Passage 4 would likely experience higher cellular stress compared to those of the parental cells, which accounts for the observed growth defect in the evolved strain. Future work in enhancing biotransformation efficiency may require effort in improving acetaldehyde detoxification capability for enhancing both ethanol utilization and biotransformation efficiency.

Overall, work up to now has examined some fundamental parameters important to characterizing a biotransformation system with results providing confirmation that both the ethanol utilization pathway and cpsADH expression systems are functional in *Escherichia coli* BL21 (DE3). This is important as both gene expression systems are driven by an autoinducible promoter specific for *E. coli*. At another level, the results obtained also signifies that the ethanol utilization pathway could regenerate NADH to help power butanone biotransformation. More importantly, the reduction of butanone to 2-butanol by cpsADH in whole cell biocatalysis is a good model system for gaining an understanding of the relative efficacy of different NADH cofactor regeneration systems. With the fundamentals of the project in place, work in the next phase of the project would revolve around the construction of a plasmid encoding glucose dehydrogenase, which is the state-of-the-art in NADH cofactor regeneration system, as well as further characterization of the cofactor regeneration system at the cellular level.

5.19. Constructing a Plasmid That Encodes the State-of-the-Art Glucose Dehydrogenase System

This segment focuses on the construction of a plasmid designed to encode the glucose dehydrogenase gene, which is the key component of the state-of-the-art NADH cofactor regeneration system converting glucose to gluconolactone with the generation of one NADH per turn of the reaction cycle. The plasmid backbone was the same as that used in carrying the two genes of the ethanol utilization pathway. Such a choice comes about due to the need to ensure a fair comparison between the ethanol utilization pathway and glucose dehydrogenase system in whole cell NADH regeneration. Specifically, factors such as antibiotic resistance marker and plasmid copy number are important for ensuring that the metabolic burden of harbouring and replicating the plasmids are similar in

E. coli strains carrying the two different plasmids. The two NADH regeneration systems also share the same autoinducible promoter, pthrC, and therefore should have similar relative expression level of pathway genes.

As the 780 base pair (bp) glucose dehydrogenase (*gdh*) gene is endogenous to *Bacillus subtilis*, appropriate primers were designed and synthesized to aid in the amplification of the target gene from *B. subtilis* BS1A1 available in the lab through colony PCR. Gel electrophoresis results indicated that the *gdh* gene could be efficiently amplified from the bacterial species. Similarly, using appropriate universal primers for the two flanking barcodes of the ethanol utilization pathway gene cassette, the 2.6 kilobases plasmid backbone could be effectively amplified from a plasmid purified through plasmid miniprep. The main difficulty encountered in the plasmid construction process comes from the inability of ligating the prefix and suffix barcodes to the amplified and purified *gdh* gene fragment. As the ligation process requires the amplified gene fragment to have specific modifications to aid in attaching a barcode, problems in primer synthesis such as the incorrect synthesis of a modified base in the second position of the primer sequence might be the key reason why there was a relative difficulty in ligating the barcodes to the amplified and treated *gdh* gene fragment. The problem was resolved by appending the appropriate barcode to the forward and reverse primers.

5.20. Evaluating the EUP Pathway with State-of-the-Art Glucose Dehydrogenase System

Glucose dehydrogenase (GDH) gene from *Bacillus subtilis* BS1A1 was successfully cloned and inserted into the plasmid backbone of the ethanol utilization pathway plasmid. Expression of GDH gene is controlled by pthrC autoinducible promoter. Except the inserted gene, all other aspects of the plasmid are the same as that for the ethanol utilization pathway, thereby, enabling a fair comparison to be made between the two NADH regeneration systems. Polymerase chain reaction successfully detected GDH gene in the plasmid via the use of universal barcode primers, and the sequence of the inserted gene was verified by Sanger sequencing. Assembled plasmid was transformed into *E. coli* BL21 (DE3) together with the p76 plasmid carrying cpsADH gene and the strain is named *E. coli* GDHcps1. Use of *E. coli* BL21 (DE3) as chassis facilitates fair comparison between the glucose dehydrogenase and ethanol utilization pathway.

To evaluate the relative efficacy of different NADH regeneration systems in enabling butanone biotransformation, resting cell biotransformation experiments were conducted. Specifically, the strain with ethanol utilization pathway (*E. coli* ethcps2) was tested against the GDH strain (*E. coli* GDHcps1) and an *E. coli* strain with cpsADH enzyme but no heterologous NADH regeneration system (*E. coli* cps2) serving as the control. Resting cells were precultivated in LB medium and 10 g/L was the concentration for both glucose and ethanol cosubstrate. Results revealed that the EUP strain delivered the highest biotransformation efficiency (yield per unit OD) amongst the strains tested, thereby, validating theoretical predictions that the EUP pathway is more efficient in regenerating NADH compared to glucose dehydrogenase (Figure 45a). Butanone utilization per unit OD data also revealed that *E. coli* strain with EUP pathway was better able to utilize butanone compared to GDH and other control strains (Figure 45b). This suggests that NADH regeneration capability is correlated with butanone utilization with the direction of causation running from a larger pool of NADH driving the uptake and utilization of butanone. Both EUP and GDH strains delivered higher biotransformation efficiency compared to the control strain where corresponding endogenous ethanol and glucose utilization pathways helped regenerated NADH. This demonstrated that both the EUP and GDH NADH regeneration systems are active and functional. More evidence supporting this view can be found in the utilization of co-substrate, glucose and ethanol. Specifically, strain with EUP consumed more ethanol compared to strain with endogenous ethanol utilization pathway (i.e., *E. coli* cps2). Similarly, strain with GDH consumed more glucose compared to strain with endogenous glucose utilization pathway only. Finally, acetate formation was also highest in cells with EUP pathway compared to those with GDH or endogenous enzymes (Figure 45d). Such a result strongly affirms that the EUP pathway is functional where acetate becomes the major byproduct in cells in the resting mode compared to the growth mode.

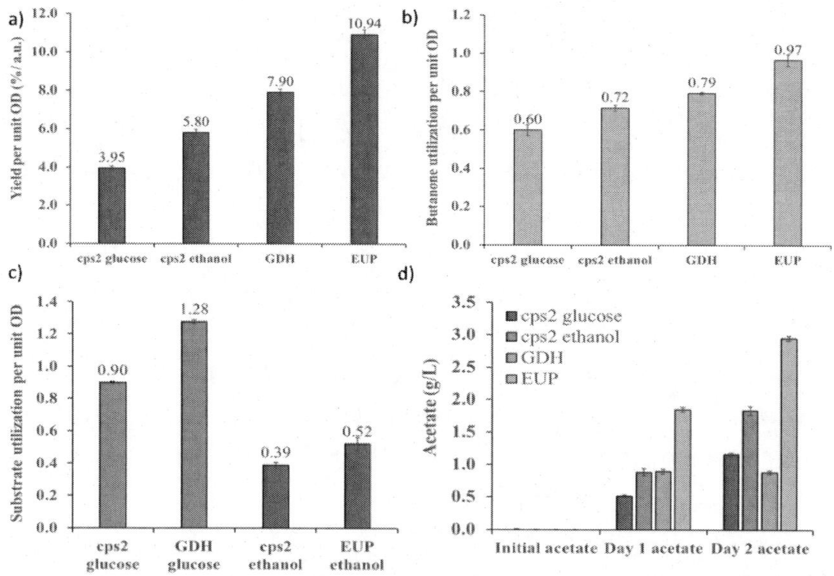

Figure 45. Comparison of NADH regeneration efficiency of different regeneration systems in strains precultivated in LB medium. a) EUP delivered the highest biotransformation efficiency compared to GDH and endogenous ethanol and glucose utilization pathways, b) Similarly, EUP also engender the highest utilization of butanone per unit OD, thereby, suggesting that NADH availability may be the driving force of butanone utilization, c) cells with either GDH or EUP utilized more of their co-substrate (glucose or ethanol) compared to cells with endogenous pathways, d) Acetate formation was highest in cells with EUP pathway, which dovetails with theoretical predictions that acetate may be the terminal product from ethanol utilization in resting cells.

Previous experiments have shown that cells precultivated in M9 + 10 g/L ethanol medium delivered higher biotransformation efficiency compared to those precultivated in LB medium. This observation is again verified in this set of experiments seeking to identify the best performing NADH regeneration system irrespective of whether glucose or ethanol served as the co-substrate (Figure 46a). One possible reason underpinning why precultivation in M9 medium would yield a better biotransformation efficiency is the expression of a set of genes that would not deplete the pool of NADH regenerated by the cofactor regeneration system. In terms of yield per unit OD, cells with EUP pathway delivered the highest biotransformation efficiency compared to cells with GDH and endogenous cofactor regeneration systems. This again reaffirms that EUP pathway is more efficient in regenerating NADH compared to GDH system. The same trend is also observed in data for butanone utilization per unit OD (Figure 46b). Specifically, cells with EUP pathway consumed more butanone compared to those with GDH, which provided a positive correlation between NADH regeneration efficiency and butanone utilization. What likely happens is that high NADH regeneration provides the driving force for butanone utilization. Co-substrate utilization, on the other hand, serves to help verify if the heterologous EUP and GDH pathways are active and functional. Observations of higher glucose and ethanol consumption in cells with heterologous GDH or EUP pathway strongly suggest that the heterologous pathways were successfully expressed and functional (Figure 46c). Finally, acetate formation was highest in cells with EUP pathway compared to those with GDH and endogenous cofactor regeneration systems (Figure 46d). This comes about due to the nature of EUP pathway that ends with acetyl-CoA. Without depletion from growth processes, acetyl-CoA may be converted to acetate and secreted out of the cell, which suggests that high ethanol consumption and acetyl-CoA formation should correlate with high extracellular acetate concentration.

Overall, this set of performance characterization tests validates theoretical predictions that EUP pathway would be more efficient in NADH regeneration than glucose dehydrogenase (GDH) system. The effect is independent of the type of growth medium used in precultivating the cells for resting cell biotransformation. Similar to prior experimentation, preculture in M9 medium seemed to activate a gene expression pattern that drains less regenerated NADH from biotransformation reaction; thereby, allowing a higher yield per unit OD to be obtained. Higher biotransformation efficiency is positively correlated with higher butanone utilization, which suggests the link that larger NADH pool serves as the driving force for butanone utilization. Acetate formation is also highest in cells with EUP pathway, which verifies that the EUP pathway that terminates in acetyl-CoA is functional. The observation also dovetails with less diversion of acetyl-CoA to growth processes in resting cells, which help enhance the flux from acetyl-CoA to acetate.

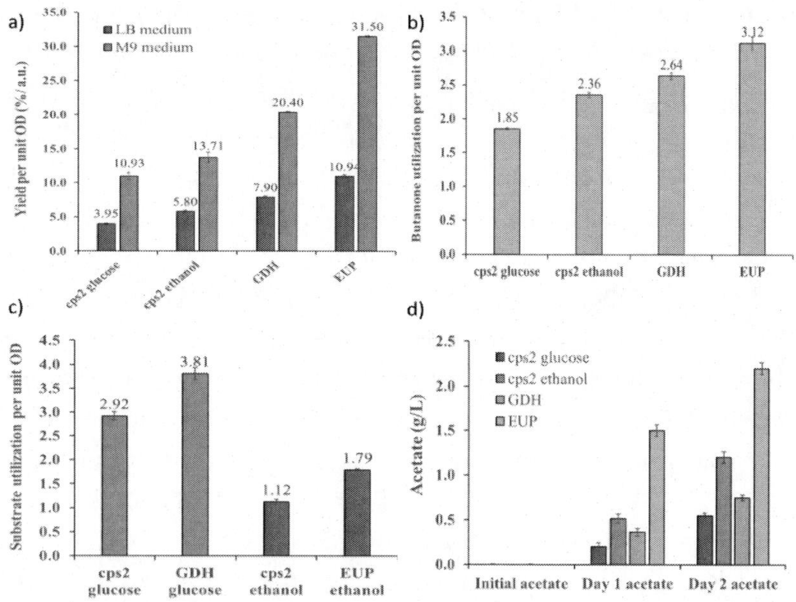

Figure 46. Comparison of NADH regeneration efficiency of different regeneration systems in cells precultivated in M9 + 10 g/L ethanol medium. a) In general, cells precultivated in M9 ethanol medium delivered higher biotransformation efficiency compared to those precultivated in LB medium. Irrespective of the precultivation medium, cells with EUP pathway delivered the highest biotransformation efficiency, b) Cells with EUP pathway also registered the highest butanone utilization per unit OD, which reaffirms the positive correlation between NADH regeneration and butanone utilization, c) Utilization of co-substrate revealed that cells with either GDH or EUP consumed more of the respective co-substrate compared to their counterparts with endogenous utilization pathway, d) Acetate formation was highest in cells with EUP pathway, thereby, affirming that the EUP pathway is functional in converting ethanol to acetyl-CoA and on to acetate in cells without growth processes depleting acetyl-CoA pool.

5.21. Titrating Ethanol Concentration to Probe for Another Driver in Biotransformation Efficiency

As a co-substrate in the biotransformation process, ethanol represents a cost that should be minimized to improve process economics. In addition, there appears to be a fundamental limit on the amount of ethanol that could be utilized by the cells irrespective of whether butanone was added to the cell suspension under anaerobic conditions. Thus, experiments were designed to evaluate the effect of different ethanol concentrations on biotransformation efficiency in *E. coli* ethcps2 under resting cell and growing cell conditions. Cells were precultivated in either LB or M9 + 10 g/L ethanol medium for resting cell and growing cell biotransformation. The strain chosen, *E. coli* ethcps2, harbours both cpsADH and ethanol utilization pathway. Results revealed a direct correlation between ethanol provision and yield per unit OD and product titer in resting cells (Figure 47 a and b). This suggests that substrate toxicity effect from ethanol was absent in the resting cell biotransformation system and 10 g/L ethanol does not exert substantial toxicity on resting cells. Besides yield per unit OD and product titer, ethanol provision was also positively correlated with ethanol utilization rate (Figure 47c) and acetate formation (data not shown). The former indicates that higher ethanol concentration provides a greater driving force for its utilization, while the later suggests that acetate formation is directly correlated with ethanol utilization, which confirms that detoxification of acetaldehyde via acetaldehyde to acetate shunt is important in addition to the conversion of acetyl-CoA to acetate through acetyl-phosphate in resting cells. Finally, butanone utilization per unit OD also shows a direct dependence on ethanol concentration (Figure 47d), which suggests that large ethanol provision drives ethanol utilization and NADH regeneration. Subsequently, a larger NADH pool in the cell would propel the uptake and utilization of butanone, which is fundamental for higher butanol formation.

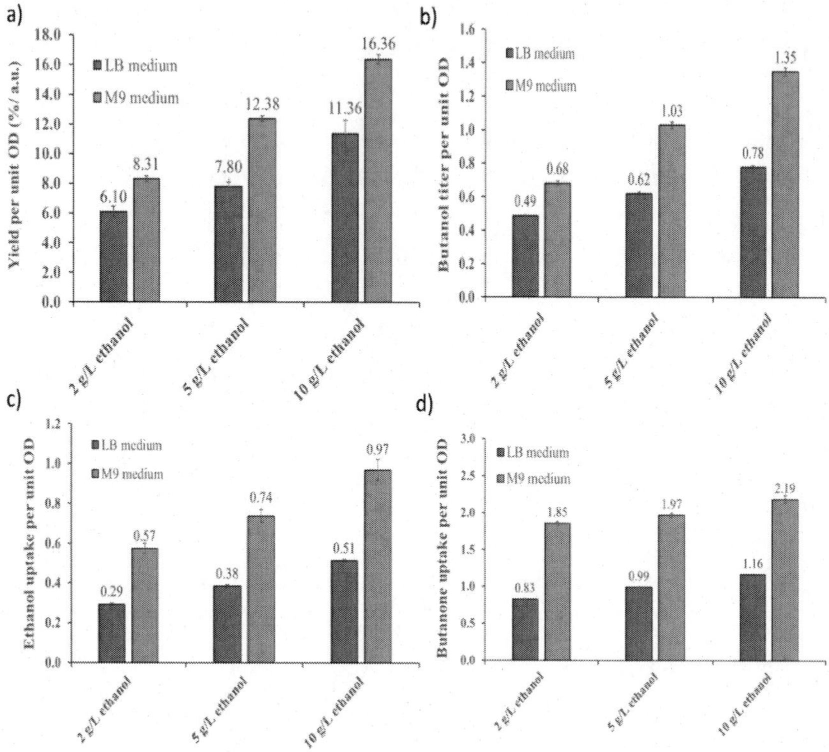

Figure 47. Effect of ethanol provision on biotransformation efficiency in resting cell biotransformation with *E. coli* ethcps2 precultivated in either LB or M9 ethanol medium. a) Positive correlation exists between yield per unit OD and ethanol provision across both medium tested, b) Similar to yield per unit OD, butanol titer also shows linear correlation with ethanol provision, c) Higher ethanol uptake was observed with ethanol provision, thereby, indicating that the EUP pathway is functional and no bottleneck exists in the pathway, d) Butanone utilization per unit OD values also show direct dependent on ethanol provision, which provide credence to the postulation that larger NADH pool and regeneration drives utilization of butanone.

Compared with data from resting cell experiment, growing cell experiment utilizing different concentrations of ethanol revealed a substrate toxicity effect from ethanol at 10 g/L ethanol. Specifically, direct correlation between ethanol provision and yield per unit OD was observed between 2 and 5 g/L ethanol, but the correlation turned negative at 10 g/L ethanol (Figure 48a). This is understandable given that growing cells are more sensitive to toxicity effect from ethanol than resting cells. At a more granular level, observation of substrate toxicity effect of ethanol in growing cell experiment is indicative of the sensitivity of *E. coli* ethcps2 to ethanol [45]. In addition, *E. coli* ethcps2 exhibited limited growth in M9 ethanol medium under anaerobic biotransformation conditions across all tested ethanol concentrations (Figure 48b), which suggests that cell growth could not be titrated by ethanol provision, and may be due to fundamental problems with the EUP pathway in the cellular context of a gene expression pattern activated by M9 medium. Specifically, there is likely a pathway bottleneck at acetaldehyde that prevents flux from flowing efficiently to acetyl-CoA. More supporting evidence for the above conjecture comes from the observation of high ethanol utilization in *E. coli* ethcps2 cultivated in M9 medium with 10 g/L ethanol (Figure 48c). Such a result attests to the fact that ethanol conversion to acetaldehyde was functioning optimally, but the conversion met a roadblock at acetaldehyde. Judging from the small amount of acetate produced during growing cell biotransformation, intracellular accumulation of acetaldehyde may be a key reason underlying poor growth of *E. coli* ethcps2 with ethanol as sole carbon source (Figure 48d). In addition, lack of acetate formation after 1 day of biotransformation suggests that the kinetics of acetaldehyde detoxification may be slow. Overall, the results point to the need for further engineering the ethanol utilization pathway to improve acetaldehyde flow to acetylCoA or acetate through the endogenous acetaldehyde to acetate shunt.

Growth medium has been shown in earlier experiments to have an important effect on overall biotransformation efficiency. Specifically, cells cultivated in M9 ethanol medium was able to deliver higher yield per unit OD compared to those grown in LB medium. This implies that gene expression patterns in cells grown in M9 ethanol medium was more in sync with biotransformation potential of the strain. Particularly, many enzymes that tap on the intracellular NADH pool may not be active, and the NADH regenerated could be used for biotransformation; thereby, generating

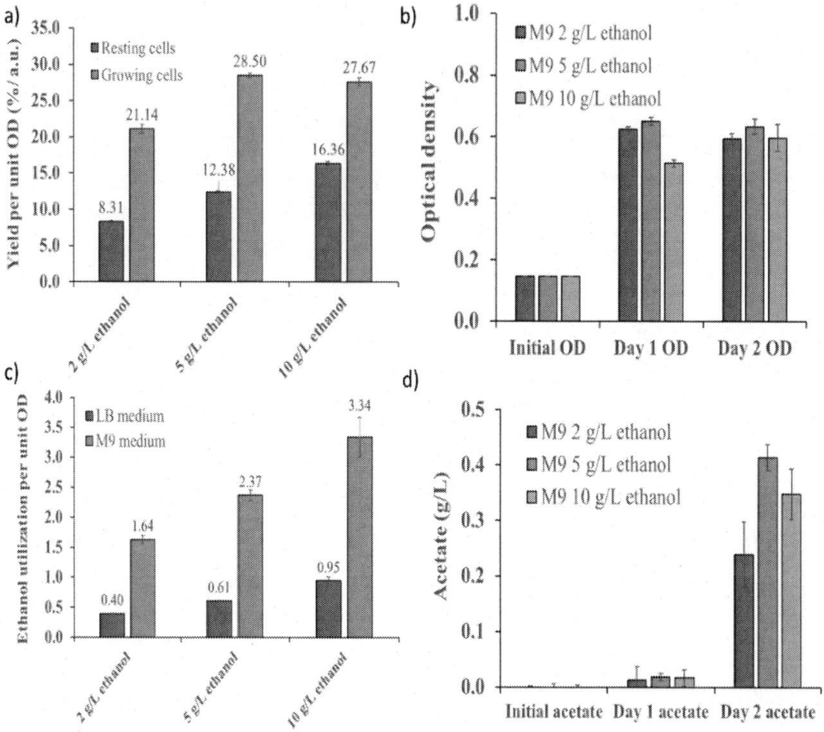

Figure 48. Ethanol toxicity effect was observed at 10 g/L ethanol in *E. coli* ethcps2 cultivated in M9 medium. a) Co-substrate toxicity effect was observed at 10 g/L ethanol in *E. coli* ethcps2 cultivated in M9 medium for growing cell biotransformation, b) Very limited growth was observed for *E. coli* ethcps2 in M9 medium across all ethanol concentration, thereby, indicating that gene expression pattern activated by growth in M9 medium is not conducive for growth with ethanol as sole carbon source, c) Ethanol utilization was higher in cells cultivated in M9 medium compared to LB medium, and showed positive correlation with ethanol provision. Such a result indicates that conversion of ethanol to acetaldehyde may not be affected, what causes toxicity may be the accumulation of acetaldehyde, d) Lack of acetate formation during growing cell biotransformation in M9 medium dovetails with earlier postulation that acetaldehyde may have accumulated intracellular with lack of conversion to acetate and acetylCoA. In addition, substrate toxicity effect was also observed at 10 g/L ethanol given poorer acetate formation at this co-substrate concentration.

higher yield per unit OD. However, cell growth was retarded in *E. coli* ethcps2 cultivated in M9 ethanol medium, which suggests that if the bottleneck in cell growth could be alleviated, better biotransformation efficiency might be achievable. To this end, results from experiments examining the effect of ethanol provision on biotransformation efficiency of *E. coli* ethcps2 cultivated or precultivated in LB medium suggest that expression of enzymes capable of detoxifying acetaldehyde may play a key role in enabling better cell growth; thereby, suggesting pointers for further engineering of the ethanol utilization pathway for cofactor regeneration.

Specifically, the central theme emanating from work done on resting and growing cells exposed to different ethanol concentrations tells a story where enzymes capable of detoxifying acetaldehyde were better expressed in LB medium compared to M9 ethanol medium. This directly leads to better cell growth in LB medium irrespective of ethanol concentrations (Figure 49a), since accumulation and toxicity from acetaldehyde may impact on growth performance of *E. coli* ethcps2 even during cultivation in LB medium. Furthermore, absence of substrate toxicity effect for ethanol provided at high concentrations confirms that cells cultivated in LB medium were better able to detoxify acetaldehyde. Theoretically, this should have led to larger amount of acetate formation. However, absence of higher acetate formation in cells grown in LB compared to M9 medium suggests that other pathways for detoxifying acetaldehyde (such as converting the aldehyde to pyruvate) may be more active in *E. coli* ethcps2 cultivated in LB medium (Figure 49b). Hence, engineering better acetaldehyde detoxification capabilities in the ethanol utilization pathway is the primary objective of future work. Two possible approaches exist in achieving this goal. One involves the expression of an aldehyde dehydrogenase to convert acetaldehyde to acetate. Another pertains to the selection of a better ada enzyme to convert acetaldehyde to acetyl-CoA. While both approaches are possible, converting acetaldehyde to acetate would be more preferable for biocatalytic systems. On the other hand, having an enzyme capable of more efficient conversion of acetaldehyde to acetyl-CoA would be more useful for metabolic engineering applications given the important role of acetyl-CoA as a cellular building block.

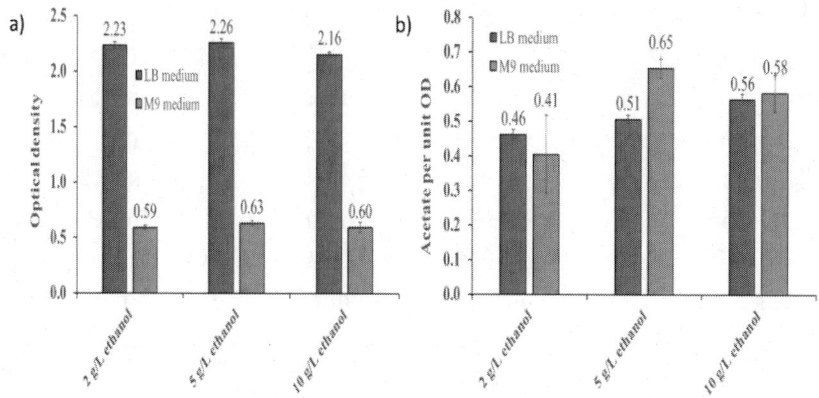

Figure 49. Growth defect of *E. coli* ethcps2 in M9 ethanol medium may be resultant from insufficient detoxification of acetaldehyde pathway intermediate. a) *E. coli* ethcps2 is able to achieve higher growth in LB medium compared to M9 medium across all ethanol concentrations tested. This suggests that growth in LB medium supplemented with ethanol is better able to activate enzymes that could detoxify accumulated acetaldehyde from ethanol utilization by EUP pathway. b) Comparable acetate formation was observed in cells cultivated in LB ethanol and M9 ethanol medium, which suggests that other pathways may be enabling the detoxification of acetaldehyde in cells grown in LB ethanol medium other than the acetaldehyde-acetate shunt. One such possibility is the formation of pyruvate from acetaldehyde.

5.22. Creating an Irreversible Flux from Ethanol by Using the Acetaldehyde to Acetate Shunt

Limited ethanol utilization in the current two gene EUP pathway motivated the search for alternative enzymes for creating an irreversible flux from ethanol to an intermediate such as acetate. One such possibility is the use of aldA, a NAD^+ dependent aldehyde dehydrogenase in *E. coli*, in converting acetaldehyde to acetate. This enzymatic conversion is irreversible and should provide an enhanced driving force for ethanol utilization. Such a construct would also help detoxify acetaldehyde and help improve cell growth in M9 ethanol medium as well as biotransformation efficiency. Construction was thus initiated for a plasmid with pthrC autoinducible promoter with aldA and adh2 as ethanol utilization pathway genes. The new plasmid is named ethWF1. The new pathway is similar in architecture with the initial EUP pathway except for a branch at acetaldehyde bringing flux to acetate instead of acetyl-CoA.

Similar to prior characterization studies, strain harbouring the new ethanol utilization pathway and cpsADH (i.e., *E. coli* ethWF1cps2) was guided through resting and growing cell experiments in LB and M9 medium.

Experiment results in cells (precultivated in LB medium) in resting cell biotransformation revealed that ethanol utilization remains limited in the new strain. This comes as a surprise, but closer examination of the experimental system and acetate production reveals that the limit may arise due to a combination of an anaerobic biotransformation system as well as lack of performance from the aldA enzyme. In terms of biotransformation efficiency, the new strain was inferior in biotransformation when fed with ethanol compared to glucose (Figure 50a). This likely arise from the lack of NADH regeneration from aldA that limited biotransformation efficiency. Butanone utilization tells a similar story where cells fed with ethanol uptake less butanone than those fed with glucose (Figure 50b). Such a result dovetails with our current understanding that NADH regeneration drives butanone uptake, which implies that lack of NADH regeneration reduces butanone uptake. Note that cells not provided with co-substrates could uptake butanone but there was no biotransformation due to lack of NADH regeneration. On the other hand, ethanol utilization level of the strain with modified EUP was similar to that of cells with original EUP and remains limited (Figure 50c). The result indicates that the initial step converting ethanol to acetaldehyde remains functional, but the second step converting acetaldehyde to acetate may suffer from an enzyme with poor kinetic properties. Confirmation of the latter comes from acetate per unit OD data that shows that the modified EUP pathway generated less acetate compared to the original EUP pathway (Figure 50d). This shows that the aldA enzyme native to *E. coli* may have poor kinetic properties, and may not play an important role detoxifying aldehyde in the Gram-negative bacterium.

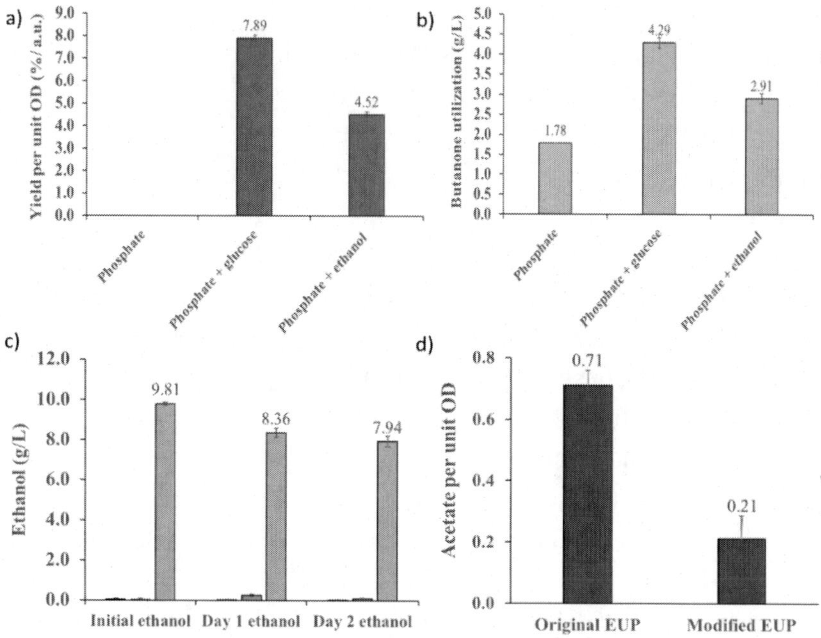

Figure 50. Biotransformation efficiency was lower for strain with modified EUP channelling acetaldehyde to acetate in resting cell biotransformation. a) Yield per unit OD was lower for cells fed with ethanol compared to glucose, signifying that the new strain could not regenerate NADH, b) Butanone utilization tells a similar story where lack of NADH regeneration reduced butanone uptake and biotransformation, c) Ethanol utilization levels remain similar to cells with original EUP, which suggests that the first reaction of the EUP pathway converting ethanol to acetaldehyde remains functional, d) Modified EUP generated less acetate per unit OD compared to original EUP in what is a surprise finding. The result suggests that aldA that converts acetaldehyde to acetate may not have favourable kinetic properties.

On the other hand, biotransformation efficiency in growing cell in LB medium exhibited the opposite trend with higher butanol yield and biotransformation efficiency for cells cultivated in ethanol compared to glucose (Figure 51a). This suggests strongly that the endogenous enzymes in *E. coli* converted the acetate end product into acetyl-CoA through possibly acetyl-CoA synthetase. This feeds into the cellular need for acetyl-CoA, and thus, set up a driving force for ethanol utilization. Cell growth in growing cell biotransformation in LB ethanol medium was however lower in cells with modified EUP pathway compared to the original EUP pathway (Figure 51b). This could be due to reduced acetyl-CoA generation from the modified EUP pathway that terminates in acetate. Ethanol utilization, on the other hand, shows that growing cells consumed more ethanol compared to resting cells (Figure 51c), which suggests that cell growth could still generate a driving force for ethanol utilization largely through the conversion of acetate to acetyl-CoA by endogenous enzymes. Finally, dismay acetate formation in cells with modified EUP fed with ethanol in LB medium casts a picture of possibly poor kinetic performance of aldA that resulted in poor acetate formation (Figure 51d).

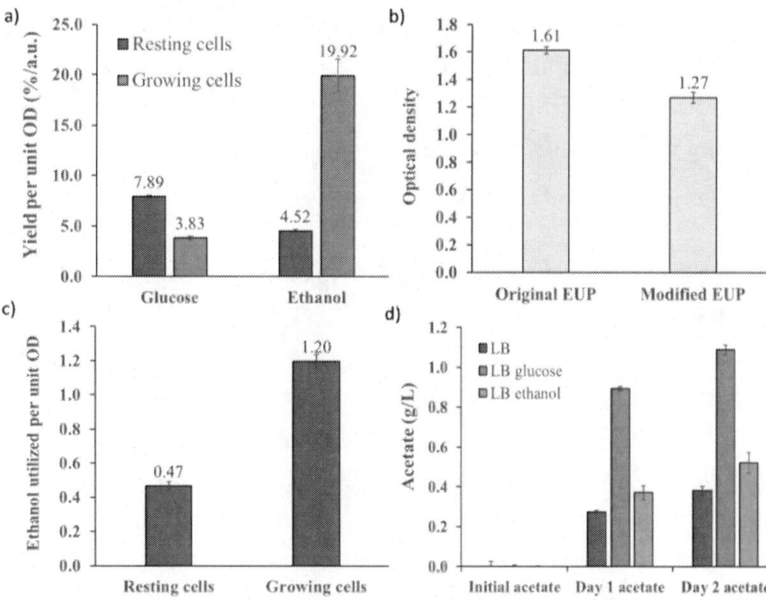

Figure 51. Cell growth could compensate for deficiency in aldA kinetic properties in cells with modified EUP in growing cell biotransformation in LB medium. a) Yield per unit OD was higher for cells with modified EUP in growing cell biotransformation with ethanol as cosubstrate, b) Cell growth was, however, lower in cells with modified EUP than those with original EUP in LB medium; thereby, suggesting that acetyl-CoA pool is an important determinant of cell growth potential, c) Ethanol utilization was higher in growing cells compared to resting cells, which reveals that cell growth could still generate a driving force through the modified EUP pathway via conversion of acetate to acetyl-CoA, d) Acetate formation was very limited in cells fed with ethanol, which is a diagnostic of poor kinetic performance of the introduced aldA enzyme.

Overall, the growing cell biotransformation system eclipsed the resting cell system in yield per unit OD, and this could be attributed to the driving force generated by cell growth that help consume acetyl-CoA, which, in turn, forces ethanol flux to flow to acetate and acetyl-CoA. Another interesting feature of the new ethanol utilization pathway showcases how heterologous genes could interact with endogenous ones in creating a driving force for ethanol utilization and NADH generation. Overall, the modified EUP pathway driving ethanol flux to acetate delivered poorer biotransformation efficiency compared to the original EUP pathway. Thus, strain with the original EUP pathway was used in the next series of experiments to examine the feasibility of using mixed substrates (glucose and ethanol) for enhancing biotransformation efficiency in resting cell biotransformation, and for decoupling cell growth and biotransformation in growing cell biotransformation.

5.23. Assessing the Feasibility of Mixed Substrate Biotransformation

Up till now, single substrate (either glucose or ethanol) was provided as co-substrate to help regenerate NADH in support of the biotransformation of butanone to 2-butanol. Thus, possibility exists of using mixed substrates to either help provide more NADH or decouple cell growth from NADH regeneration. Specifically, in the latter scenario, glucose could serve as the key energy and carbon source for biomass formation, while ethanol could be metabolized by the ethanol utilization pathway to help provide NADH for biotransformation. Performing biotransformation using such an approach would reap benefits in allowing more NADH to be generated in supporting cell growth and biotransformation; thereby, enabling higher product yield.

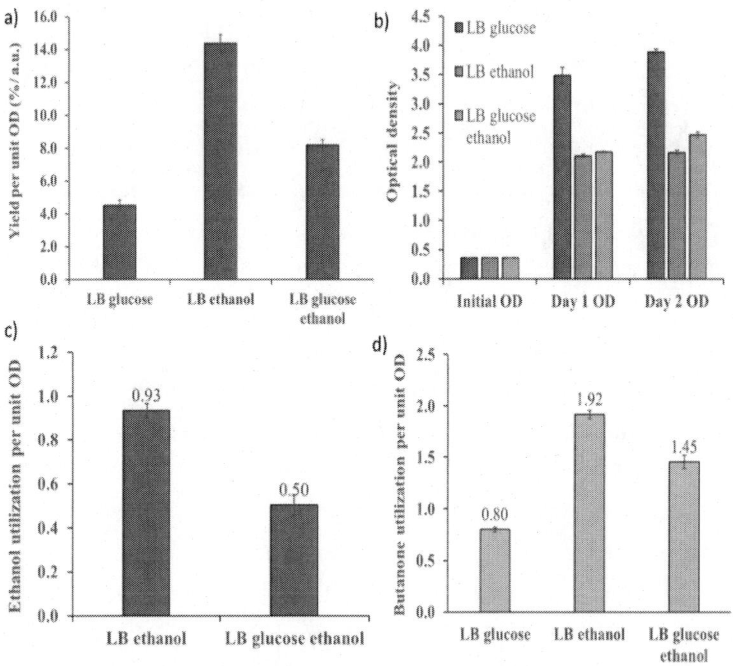

Figure 52. Mixed substrates did not elicit higher biotransformation efficiency in *E. coli* ethcps2 during growing cell biotransformation in LB medium. a) Yield per unit OD was lower for cells cultivated in LB mixed substrate compared to those in LB ethanol medium, b) Similarly, cells fed with mixed substates also achieved lower levels of growth compared to those fed with glucose only; thereby, indicating a substrate toxicity effect attributable to ethanol since such the growth defect could not be repaired with provision of glucose, c) Observation of reduced ethanol utilization in cells fed with mixed substrates relative to those fed with ethanol only suggests lower than expected expression of ethanol utilization pathway genes and glucose-dependent carbon catabolite repression may be the reason for poor ethanol utilization, d) Butanone utilization values correspond closely with those of biotransformation efficiency due to the dependence of butanone utilization on amount of NADH regenerated. Specifically, yield per unit OD values suggest that NADH regeneration was lower for cells fed with mixed substrates compared to those fed with ethanol only, and the butanone utilization values show a similar profile.

Hence, resting cell and growing cell biotransformation experiments were initiated with *E. coli* ethcps2 with both ethanol utilization pathway and cpsADH enzyme to test the hypothesis that mixed substrates (10 g/L each of glucose and ethanol) would improve biotransformation efficiency compared to cells fed with ethanol only. Results confirmed that growing cells delivered higher biotransformation efficiency compared to resting cells, due largely to the higher driving force for NADH generation arising from growth demands (data not shown). But, more importantly, growing cell biotransformation experiments in LB medium revealed the existence of carbon catabolite repression when glucose was provided to cells together with ethanol. Specifically, cells fed with mixed substrates (glucose and ethanol) delivered poorer product titer and yield per unit OD compared to those fed only with ethanol (Figure 52a). This likely arises due to the sensitivity of the pthrC autoinducible promoter to endogenous carbon catabolite repression system; thereby, resulting in lower expression levels of ethanol utilization enzymes. Such a scenario translates to lower ethanol utilization and poorer NADH regeneration compared to cells fed only with ethanol. Although glucose metabolization also generates NADH, amount produced paled in comparison to those synthesized through the ethanol utilization pathway. Cells fed with glucose and ethanol also delivered poorer growth compared to those fed with glucose (Figure 52b). This could be attributed to toxicity effect of ethanol, which is corroborated by observations of similar growth performance of cells fed with mixed substrates and those fed only with ethanol. Further evidence for existence of glucose-dependent carbon catabolite repression came from observations of lower ethanol utilization in cells fed with mixed substrates compared to those fed with ethanol only (Figure 52c). The result arose probably due to lower expression of EUP pathway genes, which suggests glucose-dependent carbon catabolite repression on the autoinducible, pthrC, promoter controlling expression of EUP pathway. Finally, butanone utilization per unit OD also showed similar trends to yield per unit OD in that cells fed with mixed substrate delivered intermediary butanone utilization compared to those fed with glucose or ethanol only (Figure 52d). Such a result manifest from the lower NADH regenerated by cells fed with mixed substrate compared to those fed with ethanol. Since NADH availability drives the utilization of butanone, reduced NADH regeneration would result in lower butanone consumption.

Similar to growing cell experiments in LB, resting cells experiment in the same medium also yielded an unexpected observation of lower biotransformation efficiency for cells fed with mixed substrate compared to those only provided with ethanol (Figure 53a). Theoretically, resting cells are endowed with glucose and ethanol utilization enzymes expressed during the growth phase and should be able to metabolize both substrates efficiently. One possible explanation for the anomalous observation stems from the postulated relative short half-life of ethanol utilization enzymes, which necessitates their renewed synthesis during biotransformation. However, with carbon catabolite repression mediated by glucose, expression levels of ethanol utilization enzymes were lower than those in cells fed with only ethanol. Thus, lower amount of ethanol was utilized which resulted in poorer NADH generation and lower yield per unit OD. Evidence for this conceptual model comes from different sets of metabolite data obtained from HPLC analysis. Firstly, glucose utilization was not significantly depressed in cells fed with mixed substrate compared to those fed with glucose only (Figure 53b). Although a reduction in glucose utilization was observed in cells fed with mixed substrates compared to those fed glucose only, the same reduction observed could be attributed to the toxicity effect of ethanol in mixed substrates. But, the most important evidence for reduced expression of EUP genes in cells fed with mixed substrates came from the substantial reduction in ethanol utilization for cells fed with mixed substrates compared to those fed with ethanol only (Figure 53c), which meant that expression level of EUP pathway genes was lower in cells fed mixed substrates compared to those fed with ethanol. Reduced expression of EUP pathway genes would likely come from the well-documented effect of glucose-dependent carbon catabolite repression on promoter controlling expression of EUP genes. Corroboration for the observation of reduced ethanol utilization in cells fed with mixed substrates relative to those fed ethanol only could also be found in acetate production data. Specifically, substantial reduction in acetate production was observed in cells fed with mixed substrates compared to those fed with ethanol only (Figure 53d). Such an observation dovetails with others that together tells the story of reduced expression of EUP pathway genes in cells fed with mixed substrates of glucose and ethanol that ultimately points to a new finding on the pthrC promoter that controls EUP expression: its sensitivity to glucose-dependent carbon catabolite repression.

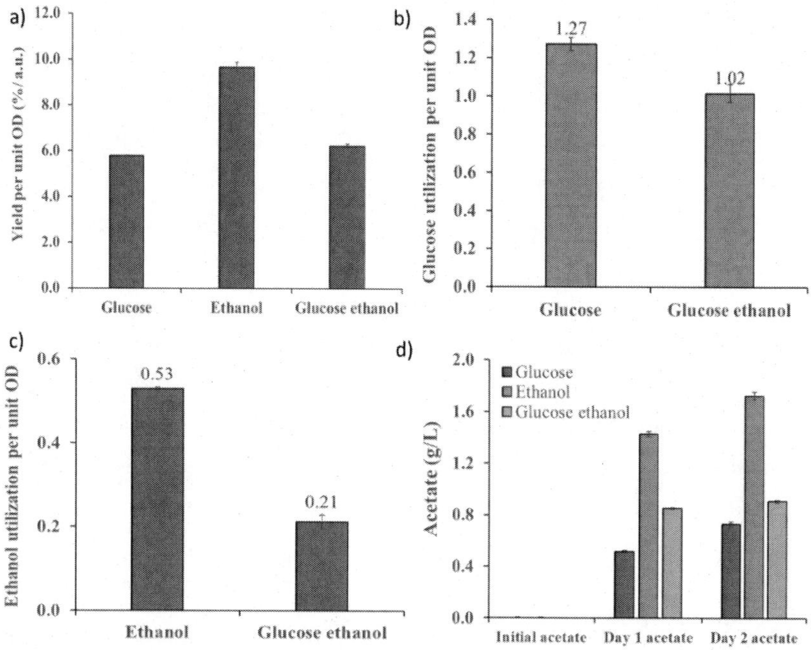

Figure 53. Mixed substrates delivered unexpected result of lower biotransformation efficiency compared to cells fed with ethanol in resting cell biotransformation with preculture in LB medium. a) Cells fed with mixed substrates delivered poorer yield per unit OD compared to those fed with ethanol only, which suggests a glucose-dependent carbon catabolite repression effect on EUP expression in resting cells, b) Glucose utilization data reveals that cells fed with mixed substrate consumed less glucose compared to those fed with glucose only in what can be attributed as ethanol toxicity on glucose uptake, c) Ethanol utilization shows the same trend as glucose utilization but with greater accentuation of effect, and provides confirmatory effect at the metabolite level that expression of EUP pathway genes was lower in cells fed with mixed substrates compared to those fed with ethanol only, d) Substantially less acetate was produced in cells fed with mixed substrates compared to those fed with ethanol due to lower abundance of EUP pathway enzymes resulting from glucose-dependent carbon catabolite repression on the pthrC promoter.

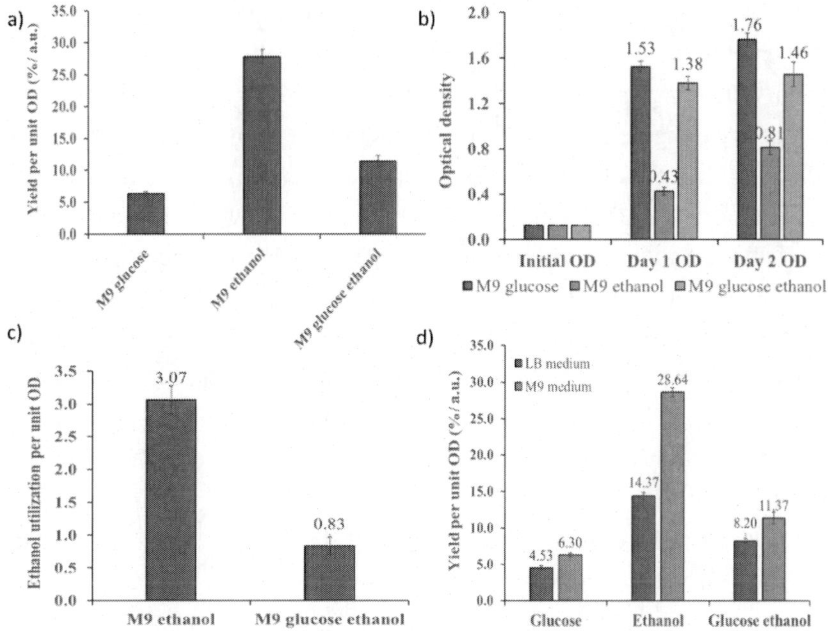

Figure 54. Growing cell biotransformation experiment in M9 medium examining the effect of mixed substrates reveals that cells fed with mixed substrates delivered lower biotransformation efficiency compared to those fed with ethanol only. a) Yield per unit OD of *E. coli* ethcps2 fed with mixed substrates in growing cell biotransformation was lower than those fed with ethanol only, b) Cells fed with mixed substrates in M9 medium could achieve higher optical density compared to those fed with ethanol only; thereby, suggesting that toxicity of ethanol is not a major issue in this biotransformation system, c) Lower ethanol utilization per unit OD in cells fed with mixed substrates relative to cells fed with ethanol only suggests that reduced expression of EUP pathway genes in mixed substrates utilization. Such a scenario could arise due to glucose-dependent carbon catabolite repression of the pthrC promoter controlling EUP pathway genes expression, d) In general, growing cells cultivated in M9 medium delivered higher biotransformation efficiency compared to those grown in LB medium across all substrate categories tests. This likely arise from the reduced activation of enzymes that could deplete NADH during growth in M9 medium.

Repeat experiments in both resting cell and growing cell mode for M9 medium confirmed the glucose dependent carbon catabolite repression effect that influenced the expression of both ethanol utilization pathway genes and cpsADH. Specifically, presence of glucose as co-substrate would reduce expression of ethanol utilization pathway genes and cpsADH that, in turn, reduced ethanol utilization and lowered biotransformation efficiency (Figure 54a). *E. coli* ethcps2 was able to achieve higher optical density during growth in mixed substrates than those fed with ethanol only during growing cell biotransformation (Figure 54b). This shows that lack of growth in cells fed only with ethanol did not arise from toxicity posed by ethanol. On the other hand, ethanol utilization values delivered confirmation that presence of glucose might have reduced EUP pathway gene expression that ultimately lead to lower ethanol utilization (Figure 54c). Finally, culture in M9 medium delivered higher biotransformation efficiency compared to LB medium in growing cell biotransformation (Figure 54d). This likely comes about through the reduced activation of enzymes that deplete NADH from the pool regenerated by EUP for aiding biotransformation.

Overall, efforts that sought to utilize the potential of mixed glucose and ethanol substrates to promote higher biotransformation efficiency or to decouple growth from biotransformation in growing cell biotransformation were met with a glucose-sensitive pthrC promoter that delivered lower expression of EUP pathway genes and cpsADH in mixed substrate environment. Hence, the concept could not be effectively tested. While at the theoretical and practical level, the current autoinducible pthrC promoter could be used for controlling the expression of EUP pathway genes, a promoter not sensitive to glucose may offer an added dimension of flexibility in utilizing the NADH regeneration capability of the EUP pathway genes in mixed substrate growing cell biotransformation.

5.24. Exploring the Use of a Constitutive Promoter for Enhancing Expression of EUP Pathway Genes

Thus far, the ethanol utilization pathway (EUP) system uses an autoinducible promoter, pthrC, for driving transcription of pathway genes, *adh2* and *ada*. Experimental results suggest that this may not be optimal given observations of a limit in amount of ethanol utilization during the experiment period. This, in turn, led me to think that expression of ethanol utilization pathway genes may not be high, and this therefore translates to a limit in amount of ethanol that could be utilized. Hence, efforts were made to express the same ethanol utilization pathway genes under a stronger promoter, and this led to studies on the effect of how a constitutive promoter (PgyrA) connected to gyrase A, a DNA helicase, would help improve expression of ethanol utilization pathway genes, and ultimately improve biotransformation efficiency through greater regeneration of NADH cofactor.

Similar to prior experiments, the same experiment system involving butanone biotransformation under both resting cells and growing cells was used in understanding whether PgyrA would be a better promoter for the plasmid carrying the ethanol utilization pathway genes given that promoter choice is a critical design parameter of a plasmid-based cofactor regeneration system. Results from resting cell and growing cell experiments conducted in LB medium confirmed that yield per unit OD was lower for cells with EUP pathway under the control of PgyrA compared to pthrC autoinducible promoter (Figure 55a and b). This is surprising since a constitutive promoter such as PgyrA should confer a higher expression level of EUP pathway genes compared to the autoinducible pthrC promoter. But, from another perspective, gyrase A is a helicase that does not require high relative abundance for optimal function in the cell. Specifically, only a small amount of gyrase A is needed, especially during DNA replication prior to cell division. Hence, while it is a constitutive promoter, PgyrA likely only engenders a low expression level for genes coupled to it. This would thus explain why biotransformation efficiency for cells with EUP under the control of PgyrA was actually poorer. Ethanol utilization values provide strong evidence suggesting that EUP pathway gene expression was lower in cells whose EUP pathway was under the control of PgyrA relative to those with EUP under the control of pthrC. Lower EUP pathway gene expression would translate into lower ethanol utilization (Figure 55c). Finally, butanone utilization was also lower for cells with EUP under the control of PgyrA compared to those whose EUP is under the control of pthrC. This arise probably from reduced expression of EUP pathway genes under the control of PgyrA that resulted in lower NADH regeneration.

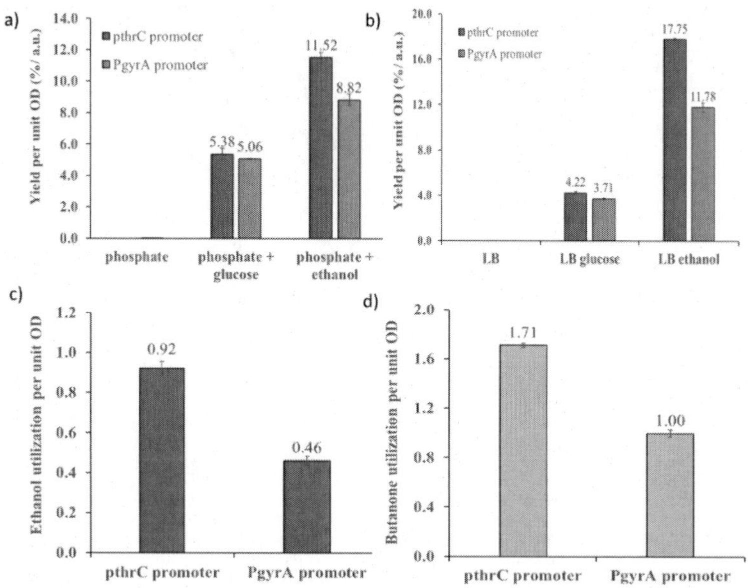

Figure 55. Expression of EUP pathway genes through constitutive PgyrA promoter resulted in lower biotransformation efficiency compared to autoinducible pthrC. a) Biotransformation efficiency was lower in cells with EUP under PgyrA control compared to the same pathway under pthrC control in resting cells precultivated in LB medium, b) the same conclusion was also obtained in growing cell experiment conducted in LB medium where cells with EUP pathway under the control of PgyrA delivered lower biotransformation efficiency, c) Lower ethanol utilization in cells whose EUP pathway is under the control of PgyrA suggests reduced expression of EUP pathway genes from PgyrA compared to pthrC, d) Butanone utilization was also lower in cells with EUP under the control of PgyrA due probably to lower EUP pathway gene expression and less NADH regeneration.

Another important observation from the experiments confirms that growing cells did provide a driving force for converting ethanol to acetyl-CoA, thereby, helping to improve biotransformation efficiency (Figure 56a). This conclusion was built upon similar observations in cells with EUP under the control of PgyrA or pthrC. Finally, the inference that PgyrA had lower expression levels compared to pthrC was also supported by observations of greater cell growth in cells with EUP under PgyrA control compared to pthrC (Figure 56b). This can only come about from lower metabolic burden of EUP expression under PgyrA control, which, in turn, meant that expression level of EUP genes were lower.

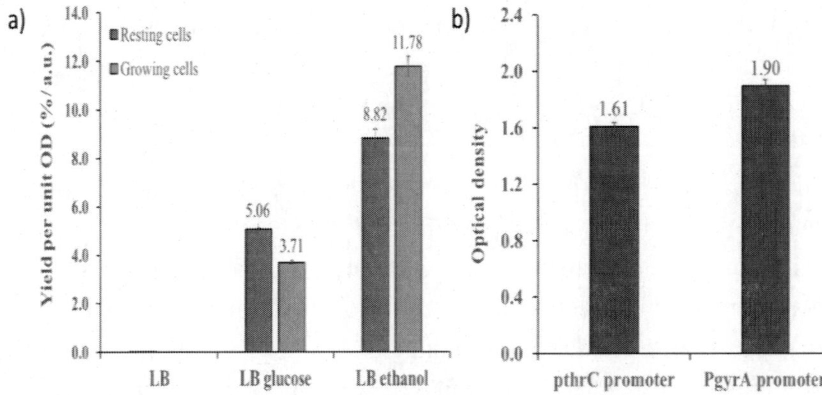

Figure 56. Confirmation that growing cells delivered higher biotransformation efficiency, and strain with EUP under PgyrA control delivered higher optical density during growth in LB medium. a) Yield per unit OD was lower for resting cells compared to growing cells of strain with EUP under control of PgyrA, b) Cells with EUP under the control of PgyrA delivered higher optical density compared to those with EUP under the control of pthrC probably due to lower expression of EUP and ensuing lower metabolic burden.

Change of growth medium from LB to M9 ethanol medium did not result in a different experiment outcome for experiments aimed to examine the relative efficiency of PgyrA and pthrC in enabling butanone biotransformation. Specifically, in resting and growing cells experiments with M9 medium, strain with PgyrA controlling EUP expression yield a lower biotransformation efficiency compared to strains with pthrC3 controlling EUP expression (Figure 57a). This again shows that while PgyrA is a constitutive promoter, its promoter strength is not higher than that of pthrC, due most probably to the evolution constrained promoter strength where gyrase A gene does not need to be of high relative abundance. More importantly, the experimental results highlight that it is not always the case that a constitutive promoter would outperform an autoinducible one. Cells with EUP under the control of PgyrA also had lower butanone utilization most probably due to the reduced regeneration of NADH from cells with less abundance of EUP pathway genes (Figure 57b). Evidence for lower expression of EUP pathway genes in cells with EUP under the control of PgyrA came from values of ethanol utilization during growing cell biotransformation. Specifically, lower ethanol utilization was observed for cells with EUP under the control of PgyrA, which strongly suggests reduced expression of EUP genes in these cells (Figure 57c). Acetate formation, on the other hand, also showed similar trend where cells with EUP under PgyrA control delivered less acetate production compared to those whose EUP is under the control of pthrC (Figure 57d). Finally, cells precultivated or cultivated in M9 ethanol medium again outperform those cultivated in LB medium (data not shown), which highlights that M9 ethanol medium might yield a gene expression profile conducive for butanone biotransformation although the specific mechanisms mediating this effect remain elusive. One possibility is that cultivation in M9 medium activated a gene expression pattern where there are less enzymes capable of depleting NADH from the pool regenerated by EUP.

Overall, attempts to improve ethanol utilization and NADH regeneration through changing the promoter controlling the expression of EUP pathway genes from pthrC to PgyrA did not achieve anticipated success. Originally, it was hoped that the constitutive PgyrA promoter would drive higher expression of EUP pathway genes that improved ethanol utilization and NADH regeneration. However, lower ethanol utilization and NADH regeneration was observed with EUP pathway came under the control of PgyrA promoter. Such a result suggests that PgyrA promoter engendered a lower expression level in *E. coli* probably due to the fact that gyrase A protein does not need to be of high relative abundance in the cell. Hence, work involving PgyrA promoter would not proceed further. Attention thus switches to the exploration of whether the endogenous promoter for adhE would be suitable for use in the EUP plasmid particularly from the perspective that the promoter may be ethanol sensitive, and thus, offer inducible control of the EUP pathway. But, first work must be done to examine if the promoter, PadhE, is sensitive to glucose-dependent carbon catabolite repression.

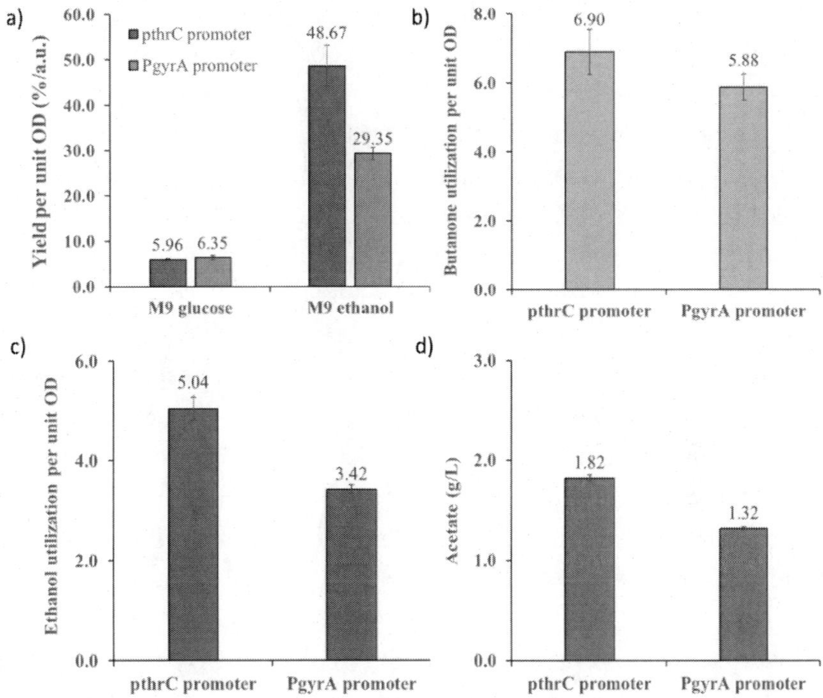

Figure 57. Examining the performance of EUP pathway under the control of PgyrA promoter in growing cell biotransformation in M9 medium. a) PgyrA promoter delivered lower biotransformation efficiency compared to cells with EUP pathway under the control of pthrC autoinducible promoter, b) Butanone utilization was also lower in cells with EUP under the control of PgyrA promoter, which suggests that there was lower regeneration of NADH in these cells, c) Strong evidence of reduced expression of EUP pathway genes under the control of PgyrA came from observations of lower ethanol utilization in these cells, d) Acetate production also showed the same trend as ethanol utilization where cells with EUP under the control of PgyrA delivered lower acetate production.

5.25. Profiling for Glucose-Dependent Carbon Catabolite Repression Effect in Adhe Promoter for Building an Enhanced Ethanol NADH Regeneration System Plasmid

An ethanol utilization pathway that relies on ethanol utilization to generate NADH for cofactor regeneration should ideally be controlled by an ethanol inducible promoter. In the absence of RNA-sequencing data that could highlight the genes induced by ethanol provision, the promoter controlling the endogenous ethanol utilization gene, adhE, is one potential candidate for cloning purposes. Prior to cloning the promoter, experiments were designed to assess if the promoter could suffer from glucose-dependent carbon catabolite repression. This fact is important given that mixed substrates comprising a mixture of glucose and ethanol could be provided to help maximize the efficiency in which different substrates could be utilized for different purposes. To help determine the characteristics of the adhE promoter, resting and growing cells experiments involving *E. coli* cps2 harbouring the cpsADH enzyme were performed in both LB and M9 medium supplemented with glucose, ethanol or glucose and ethanol.

Experimental results with LB medium indicated that biotransformation efficiency was higher in cells fed with ethanol compared to mixed substrates or glucose. The effect is independent of resting or growing cells (Figure 58a). This suggests that there is some degree of glucose- dependent carbon catabolite repression experienced by the promoter controlling the endogenous adhE gene. Cell growth was, however, higher in cells fed with mixed substrates compared to those fed with ethanol (Figure 58b). This indicated low substrate toxicity effect of ethanol on cell growth. But, the most important evidence for lower expression of adhE in cells fed with mixed substrates came from observation of lower ethanol utilization in these cells compared to those fed with ethanol only (Figure 58c). Thus, the endogenous promoter controlling expression of adhE is also sensitive to glucose regulation. Another interesting observation from the data was that acetate formation in cells exposed to ethanol was significantly lower than that for cells exposed to glucose only or mixed substrates (Figure 58d). This likely comes about due to the nature in which adhE catalyzed the catabolism of ethanol to acetyl-CoA. AdhE is an enzyme capable of performing both steps of ethanol catabolism, which meant that acetaldehyde produced could not leave the enzyme. In this case, there would be reduced production of acetate from the acetaldehyde to acetate route catalyzed by aldA and aldB aldehyde dehydrogenase native to *E. coli*. Finally, growing cells again outperform resting cells in butanone biotransformation with ethanol utilization coming from endogenous genes (Figure 58a). Yield per unit optical density of 16% in growing cells cultivated in LB + 10 g/L ethanol under anaerobic conditions revealed the potential inherent in tapping on adhE in mediating ethanol utilization and NADH regeneration. All signs point to the cloning of adhE as replacement for current ethanol utilization gene as the next step in the research programme.

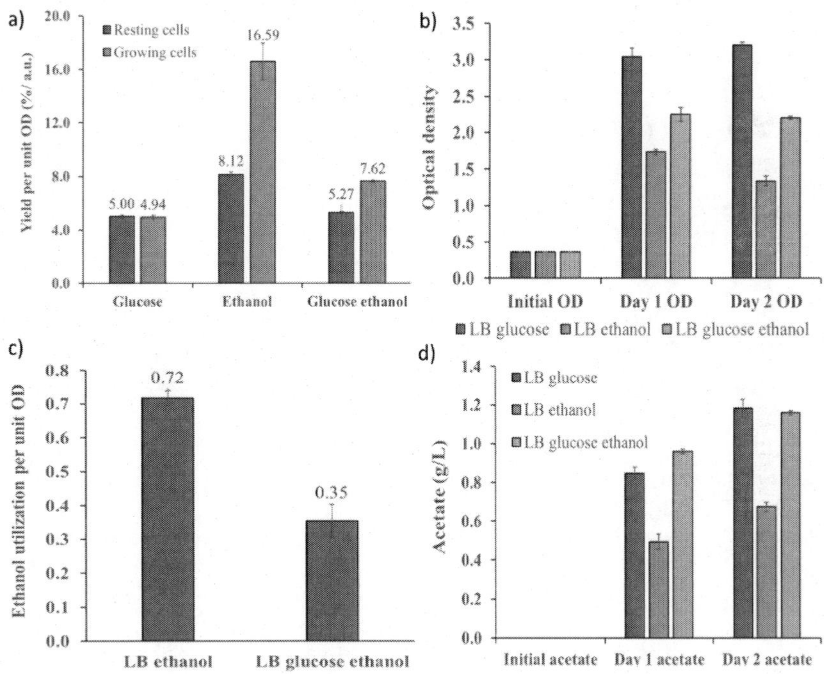

Figure 58. Mixed substrates delivered lower biotransformation efficiency in *E. coli* cps2 cells during anaerobic growing cell biotransformation in LB medium. a) Yield per unit OD was lower for cells fed with mixed substrates compared to those fed with ethanol only, with the effect independent of whether growing cells or resting cells are used for the experiment, b) Cell growth was higher for cells fed with mixed substrates compared to those fed with ethanol only, which indicates low substrate toxicity effect of ethanol on growth, c) Ethanol utilization values provided important evidence of low gene expression of adhE under mixed substrates condition compared to when ethanol was used as the main substrate, d) Acetate formation was higher in cells fed with mixed substrates compared to those fed with ethanol only.

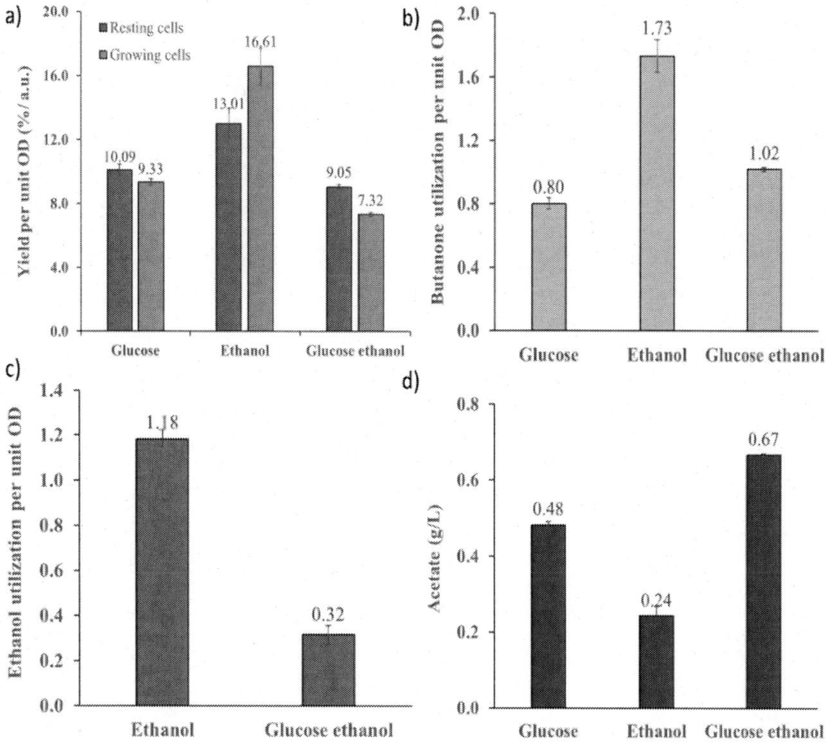

Figure 59. Mixed substrate biotransformation resulted in lower yield per unit OD in *E. coli* cps2 resting cells precultivated in M9 medium. a) Biotransformation efficiency was lower in cells fed with mixed substrates compared to those fed with ethanol only, b) Cells fed with mixed substrates also had lower butanone utilization, which corroborated that NADH regeneration efficiency in these cells were low, c) Ethanol utilization data suggests strongly that expression of endogenous adhE was lower in cells exposed to mixed substrates compared to those exposed to ethanol only, d) Acetate formation data showed that endogenous adhE was at work in ethanol utilization given the low amount of acetate produced from cells exposed to ethanol only. This arises as adhE is a dual function enzyme that converts ethanol to acetaldehyde and on to acetyl-CoA. System involving such an enzyme would have reduced acetaldehyde to acetate conversion and low acetate formation.

Replicate experiments with M9 medium corroborated the findings in LB medium. Specifically, biotransformation efficiency was highest in cells fed with ethanol compared to mixed substrates or glucose (Figure 59a). Glucose-dependent carbon catabolite repression is the most likely cause for the reduced biotransformation efficiency in cells fed with mixed substrates as the effect reduced expression of adhE in both resting and growing cells. Although resting cells are less likely to be affected by glucose-dependent carbon catabolite repression, existence of the effect in resting cells meant that half-life of adhE was relatively short and its enzyme pool requires continuous replenishment. Given that adhE catabolize ethanol to form acetyl-CoA, growing cells should ideally yield higher biotransformation efficiency compared to resting cells. Indeed, experimental findings verified this postulation (Figure 59a). Given that the endogenous adhE possess similar characteristics compared to the heterologous ethanol utilization system, it is a strong candidate for replacing the heterologous pathway in ethanol powered NADH regeneration. Lower butanone utilization values in cells fed with mixed substrates corroborated observation of lower biotransformation efficiency in these cells, and could mean that NADH regeneration was also lower. Finally, low acetate formation in *E. coli* cps2 resting cells fed ethanol revealed that endogenous adhE is the enzyme performing NADH regeneration given that its close architecture that converts ethanol to acetyl-CoA did not allow acetaldehyde intermediate to diffuse to the cytoplasm for conversion into acetate by aldA.

5.26. Expanding the Microbial Chassis in Which EUP Pathway Could Be Useful for NADH Regeneration

One important goal of this project is to develop an ethanol-based NADH regeneration system suitable for use in many prokaryotic and eukaryotic biotechnology workhorses such as *Bacillus subtilis, Escherichia coli* and *Saccharomyces cerevisiae.* Motivation for developing a generic NADH regeneration system able to operate in different microbial chassis arise from the goal of implementing the EUP pathway in many biotechnology workhorses. Hence, the cofactor regeneration system should work in Gram-positive bacteria, Gram-negative bacteria and yeast. Theoretically, the ethanol utilization pathway architecture affords it the ability to be used in a variety of Gram-positive bacteria, Gram-negative bacteria and yeast organisms. But, one concern is the presence of a toxic intermediate, acetaldehyde, in the middle of the pathway. Another concern is the potential toxicity of ethanol to the organism. But putting those concerns aside, there is nothing fundamentally wrong with the pathway that hampers its utility at NADH regeneration in different organisms. However, from the molecular biology perspective, lack of a universe promoter able to operate in all three designated species meant that some genetic finetuning is needed before the EUP cofactor regeneration system can be put to work in different chassis organisms. Notwithstanding that, there may also be a need to replace the origin of replication, ribosome binding site and transcription terminator to ready the plasmid for use in dichotomously different microorganisms on separate branches of the tree of life. In essence, as currently construed, the plasmid harbouring the EUP pathway could not be used in a "plug and play" fashion in bacterial species other than *Escherichia coli*. But, given that a lot of molecular biology experiments remains an exploration exercise, some trial and error experimentations may be helpful to confirm the above postulations. Thus, preliminary work was conducted to assess if the plasmids carrying the ethanol utilization pathway and cpsADH enzyme could be used by the replication and transcription machinery of *Bacillus subtilis* BS1A1. Resting cell and growing cell biotransformation experiments were designed for this initial foray into *B. subtilis* butanone biotransformation. One critical question asked was whether the autoinducible pthrC promoter system used on both plasmids could function in the cytoplasmic environment of *B. subtilis*. Other pertinent questions that could be answered by the experiments include whether *B. subtilis* could grow under anaerobic conditions, as well as its ability to utilize ethanol.

The first experiment involves the transformation of the biotransformation plasmid that harbours cpsADH into *B. subtilis* through inducing natural competence with 20% xylose during stationary phase growth. Formation of colonies on antibiotic selection plate meant that the transformation was successful. This strain was named *B. subtilis* cps2. Both resting cell and growing cell experiments revealed that no 2-butanol was produced, which suggests that cpsADH enzyme was not expressed given that endogenous glucose utilization should generate some NADH to push the biotransformation forward if cpsADH was present. This most likely stems from the incompatibility of the pthrC autoinducible promoter system on the plasmid with cpsADH with the endogenous cytoplasmic factors in *B. subtilis*. Specifically, the inducer molecule needed by the pthrC system could be absent from the *B. subtilis* cytoplasm, and thus, expression of cpsADH gene was not induced during exponential growth phase. Thus, future work utilizing autoinducible promoter-based plasmids in *B. subtilis* should opt for systems that uses inducers available in the endogenous cytoplasmic environment of the bacterial species. In addition, *B. subtilis* cps2 lacked endogenous enzymes for metabolizing ethanol to acetyl-CoA. This makes it an ideal strain to dissect the effect of ethanol utilization pathway on endogenous metabolism, particularly on how an influx of acetyl-CoA would affect central carbon metabolism.

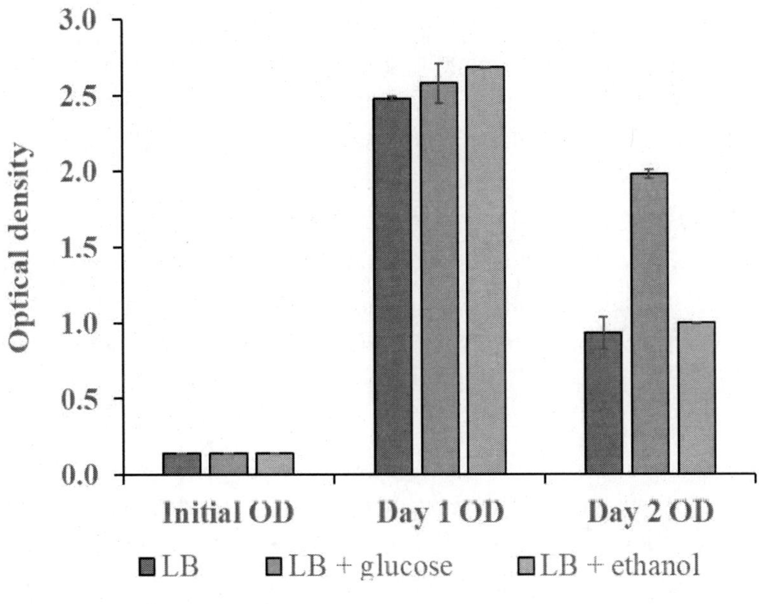

Figure 60. Time evolution of optical density of *Bacillus subtilis* cps2 cultivated in LB medium with different supplementary carbon sources for growing cell biotransformation.

Growing cell experiments revealed that *B. subtilis* cps2 could grow to an appreciable extent in LB medium with and without supplementary carbon source from glucose or ethanol (Figure 60). Comparable cell growth was observed in medium with different carbon sources, thereby, indicating that *B. subtilis* possess significant tolerance for ethanol, unlike the case for *Escherichia coli*. The data also showed that *B. subtilis* could grow under anaerobic conditions in LB medium. However, drastic decline in optical density was observed in cell population during stationary phase. Specifically, except for cells fed with glucose, those fed with LB or LB supplemented with ethanol exhibited drastic decline in optical density during stationary phase, which suggests that cannibalism might be at work in releasing cytoplasmic contents of susceptible cells to help support the remaining cell population [46, 47]. Existence of cannibalism as a cellular differentiation pathway meant that native *B. subtilis* might not be suitable as a production host for biotechnological applications. In particular, resting cell experiments that lack a full set of nutrients may induce cannibalism that decimate the cell population, thereby, resulting in poor yield.

Given that autoinducible pthrC promoter could not work in *B. subtilis*, efforts were directed towards evaluating a PgyrA constitutive promoter in *E. coli* regulating the expression of gyrase A. To do this, PgyrA promoter was coupled to the EUP pathway in a plasmid, which was transformed into *B. subtilis* BS1A1 through the natural competence technique. The resulting stain was named *B. subtilis* eth4. If transformed *B. subtilis* could grow in M9 medium supplemented with ethanol, it would mean that the EUP pathway is functional given that *B. subtilis* could not grow with ethanol as main carbon source without a heterologous pathway. Successful expression of EUP pathway genes would mean that the PgyrA promoter is functional in *B. subtilis*. However, experiment results from cell growth assays in M9 ethanol medium revealed no growth of engineered *B. subtilis* eth4 on ethanol (Figure 61); thereby, strongly suggesting that no EUP pathway genes were expressed and PgyrA promoter is not functional in the *B. subtilis* chassis. Specifically, *B. subtilis* eth4 was only able to achieve comparable growth with the wildtype strain in the same medium supplemented with 5 g/L ethanol. This suggests no additional growth on ethanol, and the EUP pathway was not active.

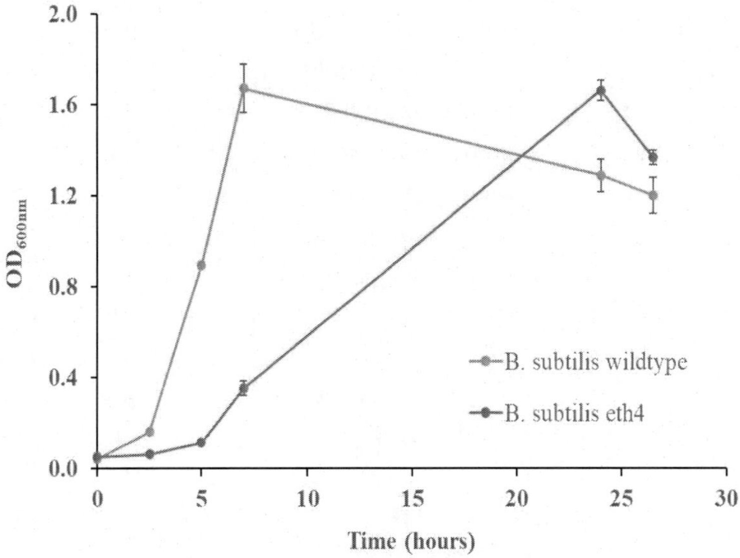

Figure 61. Assessing if *B. subtilis* eth4 with EUP under the control of *E. coli* PgyrA could grow on ethanol in M9 medium. Growth medium used was M9 medium with 5 g/L yeast extract and 5 g/L ethanol. Results suggest that *B. subtilis* eth4 could not grow on ethanol as the strain achieved the same maximal optical density as the wild-type strain in the same medium during aerobic cultivation.

Overall, the initial experiments described above paints a picture that concurs with theoretical understanding that genetic finetuning particularly in promoters is necessary to enable the current plasmid harboring EUP to be useful in another chassis organism such as *Bacillus subtilis*. Moreover, lack of growth on ethanol meant that *B. subtilis* is a useful chassis for dissecting the effect of the EUP pathway on growth, which cannot be readily done in *E. coli* due to the presence of endogenous adhE. Finally, a key lesson from the preliminary experiments is the observation of drastic decline in optical density under low nutrient conditions that presumably result from cannibalism. Such circumstances do not speak favorably for the use of *B. subtilis* in biotransformation reactions, especially resting cell biotransformation where there is a lack of a full set of nutrients.

CONCLUSION

Overall, the series of experiments conducted in this first project of the thesis validated the functionality of the ethanol utilization pathway (EUP) in enabling NADH regeneration in both single phase and two-phase biotransformation systems under anaerobic conditions in 2.0 mL HPLC vials. For example, butanone could be converted to 2-butanol in single phase biotransformation with NADH produced from ethanol conversion to acetyl-CoA. In the case of two phase biotransformation, acetophenone could be converted to 1-Phenylethanol again with the help of NADH generated by EUP pathway. Overall, the EUP pathway as implemented in *Escherichia coli*, could enable NADH regeneration when coupled with a biotransformation reaction dependent on NADH. The above coupling is tight with correlated production of transformation product (1-Phenylethanol and 2-butanol) and acetate. Aerobic and anaerobic cell growth experiments also verified that the EUP pathway could support cell growth through feeding of ethanol to acetyl-CoA.

Both resting cell and growing cell biotransformation were explored in this work to provide complementary views of the system that could inform on process performance and help identify bottlenecks that hamper, for example, biotransformation efficiency. In particular, resting cell seeks to mimic industrial conditions where cells for biotransformation are not provided with nutrients except for the biotransformation substrate and co-substrate. As construed, resting cell biotransformation is thus more cost-efficient and is less complex. On the other hand, the architecture of the ethanol utilization pathway where ethanol is converted to acetyl-CoA meant that possibility exists in using cell growth to deplete the intracellular pool of acetyl-CoA, and thereby, setting up a driving force for ethanol consumption. Hence, growing cell biotransformation are also explored in this project, and the results validated theoretical predictions. Specifically, growing cell biotransformation helped improve biotransformation efficiency compared to resting cells.

One significant finding from this series of experiment is that growing cell biotransformation works also in the case of two-phase biotransformation. Typically, the organic overlay in two-phase biotransformation is thought to induce toxicity in the cells, and thus, negatively impact on biotransformation. But, reasonably good biotransformation efficiency was obtained in two-phase acetophenone biotransformation. Ability to work in both single phase and two-phase biotransformation systems augmented EUP pathway's utility as a good candidate for NADH regeneration in both biocatalytic and metabolic engineering applications.

Substantial body of work in other fields has validated that concept of existence of differential gene expression in cells grown in different medium. In this work, two growth medium, LB and M9 + 10 g/L ethanol, were explored for their relative efficacy in enabling higher biotransformation efficiency through tuning the gene expression pattern of cells. Results revealed that cells precultivated (resting cells) or cultivated (growing cells) in M9 + 10 g/L ethanol medium persistently delivered higher biotransformation efficiency, which signifies that the gene expression pattern elicited in *E. coli* cells in M9 ethanol medium is conducive for biotransformation. This highlights an interesting area which could be further explored through the commercially available RNA-seq technology for probing the exact details of differential gene expression of cells grown in the two growth media.

To help understand the parameters that govern the biotransformation system and to validate our conceptual understanding of the process, optimization experiments were performed through tuning important parameters such as temperature, initial biotransformation substrate concentration, ethanol concentration, and promoter for EUP pathway. Specifically, temperature effect studies identified 37°C as the optimal temperature for biotransformation. Substrate concentration, on the other hand, reveals nuanced dependence that depends critically on the type of biotransformation used. Specifically, initial acetophenone concentration shows a positive correlation with biotransformation efficiency. This is in contrast to the substrate toxicity effect observed in butanone biotransformation at high substrate concentration. The divergence in system performance arises because of the protective role of the organic overlay in shielding cells from high acetophenone concentration and toxicity effect. Such protection is not available in a single phase biotransformation system such as that for butanone which is in close contact with the cells. This then explains why there is a substrate toxicity effect in butanone biotransformation system at high butanone loading.

Comparison of EUP pathway with the state-of-the-art glucose dehydrogenase system reveals that the EUP pathway is more efficient in NADH regeneration compared to glucose dehydrogenase system. This dovetails with theoretical predictions based on the structure of the two NADH regeneration pathways and amount of NADH regenerated per substrate molecule. This aspect was also indirectly validated in experiments seeking to compare the relative efficacy of endogenous glucose and ethanol utilization pathway. Results validated that endogenous ethanol utilization is more efficient in NADH regeneration and engenders higher biotransformation efficiency compared to endogenous ethanol utilization. Given that endogenous adhE has the same pathway architecture as heterologous EUP pathway, the result reaffirms ethanol utility as an efficient NADH regeneration substrate for use in biocatalysis.

Efforts to improve biotransformation efficiency saw the use of a constitutive promoter, PgyrA, in controlling the expression of EUP pathway genes. Results, however, show that the biotransformation efficiency obtained in strain with EUP under the control of PgyrA is lower than that of strain with EUP under the control of pthrC autoinducible promoter. Lower expression level of genes under PgyrA is a plausible explanation of the result and suggests that other factors may be limiting biotransformation efficiency. One possibility is the reversibility of the EUP pathway reactions that may explain why there is a fundamental constraint on ethanol utilization under anaerobic conditions. To this end, an attempt was made to modify the EUP pathway architecture by channelling acetaldehyde flux to acetate instead of acetyl-CoA. Experiment characterization of the strain with the modified EUP pathway in both resting cell and growing cell biotransformation reveals lower biotransformation efficiency compared to the original EUP pathway that ends in acetyl-CoA. Such an outcome could be due to relatively low activity of the aldA enzyme that channel flux from acetaldehyde to acetate.

Besides improving biotransformation efficiencies, efforts were also directed at exploring alternative forms of biotransformation such as mixed substrate biotransformation. In this conceptualization, glucose and ethanol would serve as the co-substrates. Ideally, glucose would power cell growth in the first phase followed by the use of ethanol in biotransformation. Hence, mixed substrate biotransformation should provide for higher biotransformation efficiency compared to what can be achieved with either glucose or ethanol alone. Results obtained, however, revealed that mixed substrate biotransformation was less efficient compared to those obtained with ethanol alone. This likely comes from the effect of glucose dependent carbon catabolite repression that reduced the expression of EUP pathway genes. Such a postulation is not without fundamental basis given that transcription of EUP pathway gene is under the control of the autoinducible pthrC promoter that may be evolutionary wired into the glucose regulon, and thus, whose transcriptional activity can be affected by the presence of glucose.

Future work in the project entails examining how the endogenous adhE enzyme compares with the two gene EUP pathway in enabling biotransformation. Given that both pathways are similar in architecture with acetyl-CoA as end-product, a one gene pathway may be preferable than the current EUP pathway in terms of reducing metabolic burden of expressing the pathway genes. Besides replacing the current EUP pathway with adhE, validating the EUP pathway at aerobic biotransformation would also expands the utility scope of the ethanol-based biotransformation system. Finally, there is a need to demonstrate the utility of EUP NADH regeneration system under aerobic biotransformation conditions to expand the application range of the system.

Chapter 3

FUTURE PROJECT A: USE OF ETHYLENE GLYCOL UTILIZATION PATHWAY FOR REGENERATING NADH IN *ESCHERICHIA COLI*

1. INTRODUCTION

Achieving redox balance has been known to help enable higher product yield in complex metabolic engineering applications where long pathways comprising steps that utilize or release redox cofactors are involved [48-50]. Such redox cofactors would typically include NAD^+, NADH, $NADP^+$ or NADPH. Less often encountered are FAD and $FADH_2$. Cases where there is an imbalance in redox environment in the cell due to the expression of heterologous redox-dependent enzymes would severely affect the oxidizing or reducing environment of cell, which have dire consequences for the efficiency of cellular processes. Hence, one of the design considerations in expressing long heterologous pathway in cells is to account for the relative balance of both the oxidized and reduced form of a cofactor such as NAD^+ and NADH. This then motivates the development of cofactor regeneration pathway or cofactor regeneration steps embedded in a long heterologous metabolic pathway.

But, the chief motivator for the development of cofactor regeneration system has been in the field of biocatalysis, which sought, as principle, to enhance the production of valueadded metabolites and products through biological routes involving enzymes either in an *in vitro* or *in vivo* environment. Cofactor regeneration becomes important in biocatalysis primarily due to the depletion of redox cofactors in cofactor dependent biotransformation reaction. Such a situation could only be solved by provision of the needed redox cofactor or brining in a redox cofactor regeneration pathway to couple with the main biotransformation route. Typically, the cofactor regeneration pathway route is chosen as it can substitute the high cost of the redox cofactor with that of the co-substrate, which if properly chosen, could be of low cost and yet enable high efficiency regeneration of needed cofactor in the pathway.

For NADH regeneration, the well-established route has been the use of glucose dehydrogenase (GDH) enzyme to regenerate one NADH through the enzymatic conversion of glucose to gluconolactone [20]. Theoretically, one NADH molecule is regenerated per molecule of glucose. This thesis has sought to improve the theoretical yield of NADH regeneration system through exploring the use of ethanol utilization pathway (EUP) in NADH regeneration in the first part of the thesis work. Results have confirmed that EUP pathway is able to engender higher efficiency cofactor regeneration compared to GDH system in both resting cell and growing cell biotransformation of butanone to 2-butanol under anaerobic conditions. Now, the thesis takes a turn to explore the tantalizing possibility of utilizing the NADH regeneration capability of ethylene glycol utilization to power cofactor regeneration for both biocatalysis and metabolic engineering.

Similar to ethanol, ethylene glycol is a C2 molecule whose biotechnological potential remains relatively underexplored. Specifically, ethylene glycol could potentially be utilized by microorganisms through its sequential conversion to glycoaldehyde and subsequently to glycolate. Prior work in Prof. Zhou's lab has demonstrated the feasibility of the approach through harnessing the endogenous fucO and aldA enzymes in *Escherichia coli* for the conversion (Figure A1). In particular, fucO is an endogenous L-1,2-propanediol oxidoreductase in *E. coli* which exhibits substrate promiscuity that could be tapped for converting ethylene glycol into glycoaldehyde. Subsequently, glycoaldehyde is detoxified into glycolate through another endogenous enzyme in *E. coli*, aldA (aldehyde dehydrogenase A). Both fucO and aldA are NAD^+ dependent and yield an NADH during the enzymatic conversion. Hence, theoretically, the ethylene glycol utilization pathway could generate two NADH molecules per molecule of the substrate. This makes the ethylene glycol utilization pathway equivalent to the EUP pathway in terms of theoretical NADH regeneration capability.

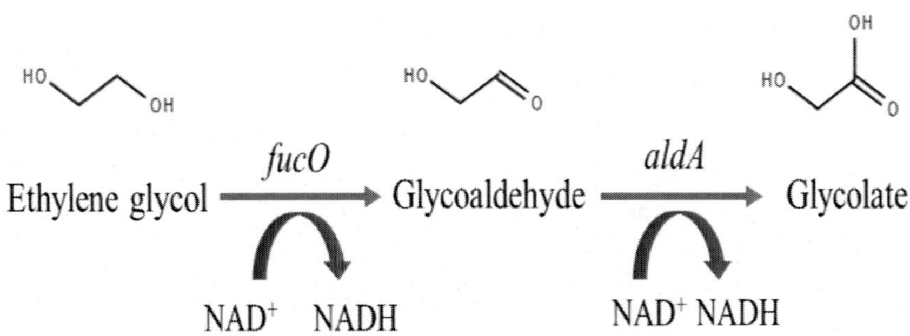

Figure A1. Pathway architecture of the ethylene glycol utilization pathway in *Escherichia coli*. The key feature of the pathway lies in its need to pass through a toxic aldehyde intermediate. Another characteristic is the ability to generate two NADH from two sequential steps of the pathway.

Cost of co-substrate is one of the key determinants of the success of a cofactor regeneration system. Glucose dehydrogenase system is dominant in cofactor regeneration largely due to the easy availability of the substrate, its low cost and lack of substrate toxicity. Ethylene glycol, on the other hand, is a relative unknown. In terms of substrate cost and availability, ethylene glycol could find a future source from recycling of plastic polyethylene terephthalate (PET) waste or through the residue from hydrolysis of lignocellulosic waste. Being a residue from recycling of abundant waste material, the cost of ethylene glycol should be reasonable even though there is lack of cost estimates for ethylene glycol from both anticipated sources. Toxicity of ethylene glycol to cells is another factor that could derail the proposed use of the substrate for cofactor regeneration, and is a point that must be checked early in the project.

Existence of a toxic intermediate in the ethylene glycol utilization pathway is another cause of concern for the utility of the pathway in cofactor regeneration. Such a situation calls for a relative high level of expression of aldA to detoxify glycoaldehyde; thereby, preventing its accumulation in the cell. To do this, the ethylene glycol utilization pathway genes could be put under the control of the well-tested autoinducible pthrC promoter from the ethanol cofactor regeneration project. Overall, experimental data from a lab colleague who first developed and tested the ethylene glycol utilization pathway revealed that the pathway could support growth of *E. coli* on up to 20 g/L of ethylene glycol under aerobic cultivation conditions (personal communication). The result suggests that toxic glycoaldehyde in the middle of the ethylene glycol utilization pathway could be detoxified effectively, and did not significantly affect cell growth on ethylene glycol.

Growing cell was shown in the ethanol cofactor regeneration project to be useful for improving ethanol consumption and NADH regeneration, and thus, biotransformation efficiency. This largely rests on the EUP pathway having a product, acetyl-CoA, that also serves as a cellular building block, whose depletion during cell growth would create a driving force for enhancing ethanol flux down the EUP pathway. Analysis of the current ethylene glycol utilization pathway indicates that a similar driving force for improving biotransformation efficiency may be harder to generate compared to the case with EUP. Specifically, the ethylene glycol utilization pathway terminates at glycolate, which is plugged into central carbon metabolism through conversion to glyoxylate via glycolate dehydrogenase (glcDEF). Glyoxylate occupies a fringe position in central carbon metabolism, and needs to condense with one molecule of acetyl-CoA to form malate that would subsequently enter the tricarboxylic acid (TCA) cycle. Hence, the analysis above suggests that any growth generated driving force for ethylene glycol utilization and NADH regeneration would likely to be small. But, it remains useful to try growing cell biotransformation for comparing the relative efficacy of resting cell and growing cell in regenerating NADH for biotransformation.

From the biocatalysis perspective, the ethylene glycol pathway as construed should work in regenerating NADH given that it does not terminate in a toxic product not tolerable by the cell. Glycolate should plug in to cellular metabolism during biocatalysis to form glyoxylate, which is not toxic to the cell. Some of the glyoxylate formed could condense with acetyl-CoA to form malate, which would ensure that there would not be any choke point in the metabolic network. Thus, ethylene glycol flux should be able to flow seamlessly towards glycolate and help regenerate NADH during resting cell biotransformation. But, one factor that must be considered is the relative stability of fucO and aldA enzymes, such as whether they need to be replenished during biotransformation. If the answer is yes, some of the ethylene glycol flux may be diverted to transcriptional and translation activity to support the energy needs of the biotransformation.

Biotransformation of butanone to 2-butanol catalysed by cpsADH would still serve as the model NADH dependent reaction for probing the utility of the ethylene glycol utilization pathway in regenerating NADH. Choice of this biotransformation reaction comes from its simplicity as well as ability to be implemented in a single phase biotransformation system. The latter allows the substrate and product to be detected in a single HPLC run which saves on time and cost. Moreover, a single phase biotransformation system is desired given that both ethylene glycol and glycolate are highly water soluble and could similarly be detected by HPLC.

Overall, what is sought after in this cofactor regeneration project is to gain application knowledge of whether ethylene glycol could emulate ethanol in delivering efficient NADH regeneration for a biotransformation reaction. Theoretically, the biotransformation reaction can be arbitrary so long as its substrate and product do not pose significant toxicity to the cell. But, what lies deeper is the scientific understanding that could be gleaned from using ethylene glycol utilization creatively to support a biotransformation reaction at the whole-cell level. Knowledge that could be gained include whether flux from ethylene glycol could flow efficiently to other metabolites in the cell, especially those in central carbon metabolism, and the efficiency in which glycoaldehyde would be detoxified by the cell in allowing the pathway to continue churning NADH.

2. Theoretical Underpinnings of the Project

As envisioned, the ethylene glycol utilization pathway converts ethylene glycol to glycoaldehyde and glycolate through the combined action of fucO and aldA. Both enzymes are native to *Escherichia coli*, but have relatively low expression levels based on RNA-seq data obtained for *E. coli* K-12 cultivated in minimal medium (data not shown). Hence, both enzymes need to be overexpressed *in vivo* in order to constitute a functional ethylene glycol utilization pathway. Such an approach has been demonstrated by a lab colleague working on converting ethylene glycol into biomass.

Examining the pathway in greater detail would provide the uninitiated with a view of the awe of evolutionary biochemistry. Specifically, while aldA is an aldehyde dehydrogenase that may be able to handle glycoaldehyde given its small size as a C2 molecule, it may require a greater leap of imagination to envision using a 1,2-propanediol oxidoreductase for performing the first step of the pathway in converting ethylene glycol to glycoladehyde in what is essentially a dehydrogenation reaction. In terms of dimension, ethylene glycol is a smaller molecule compared to propanediol, and thus, should be able to fit the active site of fucO. But, more importantly, ethylene glycol and 1,2-propanediol show structural similarity in the hydroxyl groups on the two adjacent carbon atoms in the molecule's backbone. This suggests possibility in using fucO for converting ethylene glycol into glycoaldehyde, which was validated in experiments carried out by my lab colleague. Such a result suggests that many enzymes could potentially exhibit substrate promiscuity, and efforts should be directed to examine if a particular enzyme could work on a new substrate if the active site of the enzyme could accommodate the new substrate.

The whole-cell biocatalysis system designed for this project utilizes the ethylene glycol utilization pathway to regenerate NADH for the NADH-dependent biotransformation of butanone to 2-butanol by cpsADH enzyme. Both pathways are encoded by genes on separate plasmids, which are compatible with each other in terms of replication origins and antibiotic resistance marker in order for both plasmids to be stably maintained in the *E. coli* host cells. Relatively high promoter strength autoinducible promoter, pthrC, would be used to control the transcription of ethylene glycol utilization pathway genes as well as cpsADH. Enzymes of both pathways would be transcribed and translated in the cytoplasm where the final enzyme product will mediate both coupled reactions with diffusion of intermediates and cofactors the only factor limiting the productivity and kinetics of the process. NADH serves to provide the link between the ethylene glycol utilization pathway and cpsADH mediated biotransformation (Figure A2).

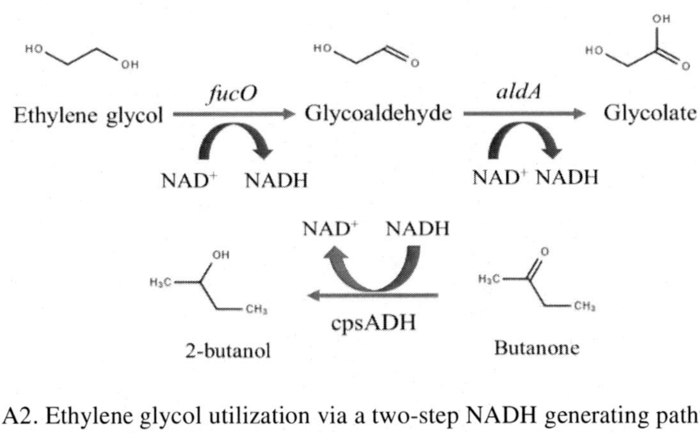

Figure A2. Ethylene glycol utilization via a two-step NADH generating pathway is coupled to the NADH dependent biotransformation of butanone to 2-butanol. All enzymes are in *E. coli* cytosol, with NADH coupling the two reactions together. Diffusion of enzymes, metabolites and cofactors are the limiting factor for the system.

Since the substrates and products are water soluble, the biotransformation system chosen for this project is a single-phase biotransformation system (Figure A3). In such a system, ethylene glycol would be taken up by the cells and be converted by the sequential reactions of fucO and aldA to form glycolate. NADH would be generated in the process and would diffuse to cpsADH enzyme for converting the butanone taken up by the cell to form 2butanol. As a terminal product, 2-butanol would diffuse out of the cell or be transported out extracellularly and be detected by HPLC. Glycolate, on the other hand, may be further converted by endogenous enzymes in *E. coli* to form glyoxylate. Thus, glycolate may not be secreted out of the cell as an extracellular product. Similarly, very little if any glyoxylate would be detected. Hence, monitoring the progress of the reaction system and diagnosing any complications will have to rely on measuring the concentration changes of ethylene glycol, butanone, 2-butanol and acetate.

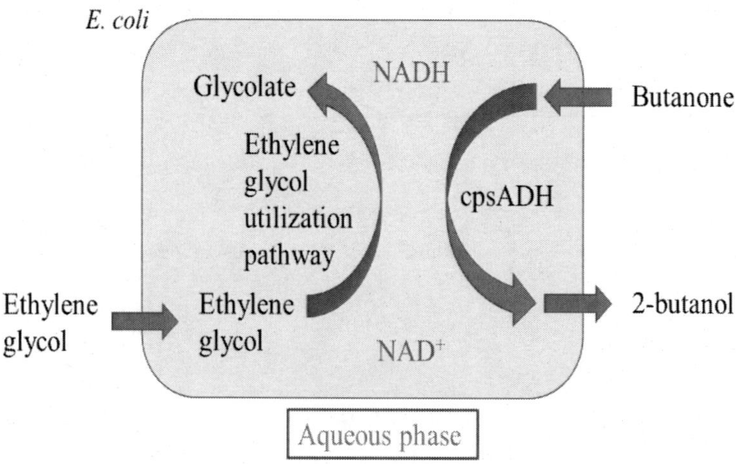

Figure A3. Schematics showing the movement of substrates and products across the cell envelop of *E. coli* during single phase whole-cell biotransformation of butanone to 2-butanol supported by NADH regenerated from ethylene glycol utilization. Integration of glycolate into central carbon metabolism meant that it is unlikely that glycolate would be present in the extracellular space in sufficient quantities for it to be detected by high performance liquid chromatography.

Resting cell and growing cell biotransformation as described in the EUP cofactor regeneration project would be carried out to assess the efficacy of the ethylene glycol utilization pathway in regenerating NADH to support the biotransformation reaction. In particular, resting cell biotransformation where no growth medium would be used is useful for assessing the inherent biotransformation potential and cofactor regeneration capability of the system. In addition, the conditions used in resting cell biotransformation resembles those used in biocatalytic biotransformation used in the industry.

Growing cell biotransformation, on the other hand, uses cell growth to generate a driving force for enhancing the utilization of ethylene glycol to help generate more NADH, which would drive the main biotransformation reaction. As mentioned, the strength and extent of the driving force generated by growth processes depends intricately on how well the ethylene glycol utilization pathway integrates with central carbon metabolism. As an example, if the metabolite that integrates with central carbon metabolism is acetyl-CoA, the driving force for biotransformation from growth processes would be significant given the experimental data elucidated in the EUP cofactor regeneration project. However, in the case of ethylene glycol utilization, the two-step pathway ends in glycolate, which could be converted into glyoxylate by endogenous glycolate dehydrogenase. Without external source of carbon flux flowing to acetyl-CoA, glyoxylate would be starved of acetyl-CoA for condensation to form malate that would help turn the tricarboxylic acid cycle. Hence, depending on the amount of acetyl-CoA available in the cell, the driving force generated by growth processes by pulling the flux from glyoxylate may not be great. Experiments would thus fill the gap to assess if the endogenous pool of acetyl-CoA in the cell is large enough to help glyoxylate to condense into malate to drive growth processes, and pull flux down ethylene glycol utilization pathway to generate NADH for supporting biotransformation. One way to circumvent the constraints on acetylCoA availability would be to supply acetate for conversion into acetyl-CoA through acetylCoA synthetase enzyme native to *E. coli*, and this represents an interesting avenue for exploration in this project.

Similar to the ethanol cofactor regeneration project, growth medium for the precultivation of resting cells or the cultivation of growing cells could also play an important role in affecting biotransformation efficiency. Hence, this project would still explore the role of LB and M9 medium in activating gene expression pattern in cells that may augment biotransformation. Such augmentation may be the reduced expression of enzymes that utilizes NADH, thereby, leaving NADH regenerated by ethylene glycol utilization for supporting biotransformation. M9 medium was shown in the ethanol cofactor regeneration project to afford higher biotransformation efficiency compared to LB medium in both resting and growing cell biotransformation due probably to reduced expression of enzymes that consume NADH. Experiments in this project would help confirm the effect and provide understanding on whether the phenomenon is generic for all heterologous pathways. If M9 medium does improve biotransformation efficiency in NADH-dependent biotransformation, it may be recommended as the medium of choice for such biotransformation, and this could become one of the major findings from the project. Whether LB or M9 medium, supplementing the medium with glucose would serve as the positive control given that *E. coli* possess endogenous glucose utilization pathway capable of generating NADH for powering butanone biotransformation.

Metrics for assessing biotransformation efficiency in this project is similar to those used in the ethanol cofactor regeneration project. Chief amongst which is yield per unit OD, which provides a basis for comparing results from resting cell and growing cell biotransformation, where the cell density loaded for biotransformation reaction are different. Similarly, yield per unit OD is also a useful parameter for understanding relative biotransformation efficiency of cells precultivated or cultivated in LB or M9 medium. Yield per unit OD is defined as per follows:

$$Yield\ per\ unit\ OD = \frac{butanol\ conc}{Initial\ butanone\ conc} \times \frac{100}{Final\ optical\ density}$$

3. Literature Review

Implementing biocatalytic systems that require redox cofactors such as NADH or NADPH in microbial cells typically require a cofactor regeneration system [14, 23]. Such systems avoid the need for provision of stoichiometric amounts of expensive redox cofactor [51], and switches the cost of the process to that of the co-substrate that regenerates the cofactor using an enzymatic reaction. In whole-cell biocatalytic systems with cofactor regeneration, the main biotransformation reaction consumes the redox cofactor, which is regenerated by a coupled second reaction that consumes a co-substrate such as glucose to produce the redox cofactor. Doing this way allows a cheap co-substrate to replace a high-cost cofactor that would otherwise be needed in stoichiometric provision of cofactor in a reaction. Choice of co-substrate is therefore important. Besides price, simplicity of reaction as well as whether the reaction product of cofactor regeneration could complicate downstream processing are also important considerations. This review focuses on NADH cofactor regeneration, even though there are many other cofactor regeneration systems available such as those for NADPH [24, 52, 53] and ATP [54, 55]. Currently, glucose dehydrogenase is the most popular NADH regeneration system for biocatalytic reaction in *E. coli*. But, cofactor regeneration systems with better atom economy, simpler reaction setup and higher productivities are still sought after [56].

Conventionally, single enzyme capable of regenerating a particular cofactor are investigated for use in cofactor regeneration, but more recently, the focus has shifted to the use of pathways for regenerating cofactors. In the latter approach, one or more enzymes that participate in the pathway are capable of regenerating a specific cofactor. In general, pathwaybased cofactor regeneration would be able to regenerate more cofactors per turn of the pathway compared to single enzyme cofactor regeneration. One approach in this direction is in the use of native pathways present in central carbon metabolism. The other approach is in the design, construction and implementation of new heterologous pathways, typically centered on utilizing new unconventional substrates, even though modified improved version of endogenous pathways have also been used [57]. In this vein, exploring the cofactor regeneration capability of ethylene glycol utilization pathway would form the focus of this segment of the thesis work.

Ethylene glycol utilization pathway links ethylene glycol to glycolate. This route was first discovered in *E. coli* when mutants were discovered to grow on propylene glycol [8]. These mutants could also grow on ethylene glycol by first converting it to glycoaldehyde, which was subsequently converted to glycolate that enters central carbon metabolism. In this incarnation of the ethylene glycol utilization pathway, fucO (an L-1,2-propandiol oxidoreductase native to *E. coli*) would catalyze the first step with regeneration of one NADH. This is followed by conversion of the glycoaldehyde produced in the first step to glycolate through the action of aldA (an aldehyde dehydrogenase native to *E. coli*), which regenerates another NADH. Hence, the two-step pathway regenerates 2 NADH per turn of the cycle. Expression levels of endogenous fucO and aldA are low in *E. coli*, and they have to be over-expressed for the purpose of constructing a cofactor regeneration system. Currently, there are no reports detailing the utilization of NADH generated by the ethylene glycol utilisation pathway for cofactor regeneration.

In terms of enzyme systems propelling the ethylene glycol utilisation pathway, there has been work done on identifying alternative enzymes capable of catalyzing the critical ethylene glycol to glycoaldehyde step. One possibility is the alcohol dehydrogenase from *Gluconobacter oxydans*, which has exhibited highly selective behaviour towards terminal hydroxyl group of varied aliphatic and aromatic diols, with no activity towards sec-alcohols [58]. This enzyme offers an option to modify the proposed ethylene glycol utilisation pathway in future work once feasibility of the pathway is successfully demonstrated.

NADH can be regenerated by a couple of enzymatic systems such as NADH oxidase (NOX) [58], glycerol dehydrogenase [13, 59], glucose dehydrogenase [60], ethanol utilization, and, in this work, ethylene glycol utilisation. Besides expanding the types of regeneration system available for the community, work on new cofactor regeneration systems could also offer an opportunity for the system to be integrated in metabolic engineering application. Overall, this literature review has arrived at the conclusion that ethylene glycol utilisation pathway's potential in providing NADH in support of a biotransformation reaction catalyzed by an oxidoreductase has not be explored, which presents an opportunity for this work. Discovery of an alternative highly selective alcohol dehydrogenase from *Gluconobacter oxydans* that could replace fucO provided more options for improving pathway efficiency and NADH regeneration in future incarnations of the pathway.

4. Materials and Methods

4.1. Bacterial Strains Used

E. coli BL21 (DE3) wildtype strain (New England Biolabs) was chosen for this work.

4.2. Chemicals and Growth Medium Used

Ethanol, ethylene glycol and glucose were purchased from Sigma-Aldrich and used without further purification. Composition of LB Miller medium is (g/L): Tryptone, 10.0; Yeast extract, 5.0; NaCl, 10.0. Composition of M9 medium with 1 g/L yeast extract is (g/L): K_2HPO_4, 6.8; KH_2PO_4, 3.0; NH_4Cl, 1.0; NaCl; 2.0; Yeast extract, 1.0.

4.3. Growth Assessment Test for Ethylene Glycol as Growth Substrate

E. coli BL21 (DE3) wildtype strain was chosen for assessing whether endogenous enzymes in *E. coli* could facilitate the utilization of ethylene glycol. Glycerol stock culture of the wildtype *E. coli* strain served as inoculum for a seed culture comprising 10 mL of LB medium in a 125 mL glass shake flask. The seed culture was cultivated at 37°C and 225 rpm for 16 hrs, after which 0.25 mL of seed culture served as inoculum for 25 mL of M9 + 1 g/L yeast extract medium (M9 base medium) in a 125 mL glass shake flask. Similarly, the seed culture also served as inoculum for M9 base medium supplemented with different carbon sources such as 5 g/L ethylene glycol, 5 g/L ethanol, and 5 g/L glucose. The experiment cultures were cultivated at 37°C and 225 rpm. The incubator used was N-Biotek NB-205 Incubator-shaker.

At appropriate time points, aliquots were withdrawn from the experiment culture for both optical density and HPLC analysis. Optical density was measured with an UV-visible spectrophotometer at 600 nm (Ambersham Biosciences). Samples were diluted with deionized water if the absorbance exceeded 1. For HPLC analysis, 0.5 mL of sample was centrifuged at 20000xg for 2 mins, 200 μL of the supernatant was carefully pipetted out and filtered through a 0.22 μm nylon filter into a 300 μL plastic HPLC vial for analysis.

4.4. HPLC Analysis Conditions

Ethanol, glucose and ethylene glycol were the target metabolites for this HPLC analysis. High performance liquid chromatography was conducted with an Agilent 1260 series instrument equipped with a Bio-rad Aminex HPX-87H column. 5 μL of sample was injected into the column and eluted with 5 mM sulphuric acid at 0.7 mL/min. Isocratic elution was used and the column was maintained at 50°C. Analytes were detected by a refractive index detector.

4.5. RNA-Seq Data Analysis

RNA-seq transcriptome analysis was called upon to assess the plausibility of certain reactions in *E. coli* through providing the relative abundance of different genes in the genome of the bacterium. To this end, the RNA-seq dataset of *E. coli* K-12 cultivated in minimal medium till stationary phase (Array Express accession number: E-GEOD-77325) was downloaded from Array Express and processed by an in-house MATLAB RNA-seq transcriptome processing software. Briefly, the approach entails the reading of a sequenced read, and aligning with all genes in the genome of *E. coli* K-12. If the sequenced read aligns with a gene, the expression count of the gene would be increased by one. In total, 3.5 million reads were processed to yield the final transcriptional picture of *E. coli* K-12 in minimal medium at stationary phase.

5. Preliminary Findings

Initial efforts in this project comes from the building of the analytical capability to detect all the metabolites, substrates and products through ion exchange HPLC analysis. Previous work on ethanol cofactor regeneration has established the detection of butanone, 2butanol and acetate by HPLC analysis. Hence, the remaining work in this segment involves the establishment of the detection method for ethylene glycol and glycolate by the same HPLC analysis method in order to afford facile and fast sample analysis for work in this project.

Analytical conditions used are as follows. The column used is a Bio-rad HPX-87H ionexchange cum reversed phase column. Mobile phase is 5 mM sulphuric acid eluted isocratically. Column temperature was maintained at 50°C and flowrate during analysis was 0.7 mL/min that helped generated a column pressure of ~88 bar. Calibration standards were prepared for ethylene glycol and glycolate in the concentration range of 0, 0.5, 1, 5, 20 g/L from their respective reagent grade chemical purchased from Sigma-Aldrich, and which were dissolved in HPLC grade water.

Samples were filtered with 0.22 μm nylon filter prior to analysis. Injection volume for each calibration standard was 5 μL. Detection was via a refractive index detector on the Agilent 1260 HPLC instrument.

Results indicated HPLC augmented with refractive index detector could reliably detect ethylene glycol and glycolate. Specifically, the retention time of ethylene glycol was determined to be 13.9 mins and that for glycolate was 10.6 mins (Figure A4a and A4b).

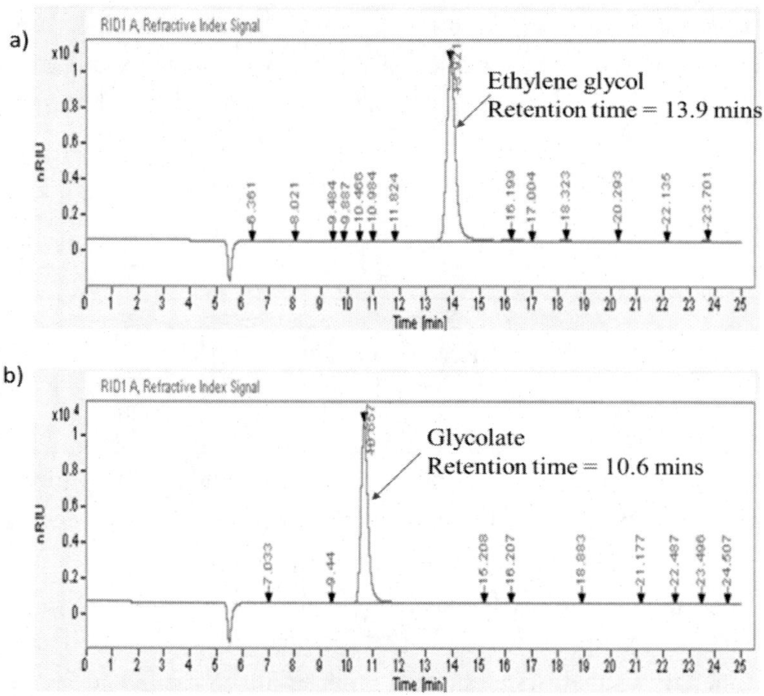

Figure A4. Chromatograms revealing the detection of ethylene glycol and glycolate with have distinct retention time that are well-separated from each other. This affords their separation and detection in a real biotransformation sample. a) Refractive index trace of chromatogram for sample with ethylene glycol, b) Refractive index trace of chromatogram for sample with glycolate.

Calibration curves for ethylene glycol and glycolate over the concentration range from 0.5 to 20g/L were linear with a high correlation coefficient between data points and best fitting line (Figure A5a and A5b). With the calibration curves for the two metabolites of the ethylene glycol utilization pathway, the analytical method for detecting the two compounds has been set-up and validated to help support biotransformation experimentation.

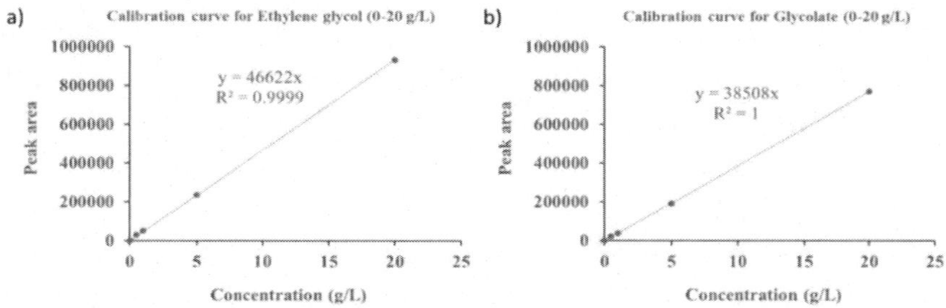

Figure A5. Calibration curves for ethylene glycol and glycolate were linear within the tested concentration range from 0 to 20 g/L and had good correlation coefficients. a) Calibration curve for ethylene glycol, b) Calibration curve for glycolate.

Ideally, a single HPLC run should be able to yield concentrations of all analytes in the single-phase biotransformation system designed in this project. To do this, each of the analyte must be well-separated from each other within a reasonable time in a single run. Table A1 lists the retention time of all analytes pertinent to this project. It could be seen that except for acetate (13.2 mins) and ethylene glycol (13.9 mins), all other analytes including glucose can be wellseparated from each other in a reasonable timeframe of 30 mins. As acetate may not be a major metabolite in cells fed with ethylene glycol, it may not interfere significantly with the HPLC analysis.

Table A1. Retention times of different analytes in the biotransformation system Compound Retention time (mins)

Glucose	7.7
Glycolate	10.6
Acetate	13.2
Ethylene glycol	13.9
Butanone	25.1
2-butanol	26.9

With the fundamentals of analytical technique in place, the initial part of the project subsequently branched off to assess if *E. coli* wildtype cells could grow on ethylene glycol and glycolate. Answers to such questions offer the basis on which future engineering of the organism could be made. Specifically, the two key enzymes in the ethylene glycol utilization pathway, fucO and aldA, are both endogenous enzymes in *E. coli*. Hence, it is important to assess if *E. coli* wildtype cells could grow on ethylene glycol with glycolate product entering endogenous metabolism through conversion to glyoxylate. Inability to grow on ethylene glycol would mean that endogenous expression of fucO and aldA are low, and therefore, argues for over-expression of the pathway in *E. coli* to mediate the utilization of ethylene glycol. From the theoretical perspective, analysis of RNA-seq transcriptome data of *E. coli* cultivated in minimal medium at 37°C revealed low expression level of fucO in *E. coli* (Table A2). Specifically, the expression count of fucO was only 8 compared to 242 for aldA. For reference, the expression level of a housekeeping enzyme with moderate expression level, gyrase A, was 574. This shows that ethylene glycol is unlikely to be metabolized by endogenous fucO given the low expression level of the enzyme.

Table A2. Expression level of different enzymes in *E. coli* cultivated in minimal medium at 37°C as determined by RNA-seq analysis

Gene	Function	Expression count
	L-1,2-propanediol	
fucO	oxidoreductase	8
aldA	aldehyde dehydrogenase	242
gyrA	gyrase subunit A	574

With the theoretical predictions in hand, growth experiments were conducted to evaluate if the predictions made using RNA-seq data for *E. coli* in minimal medium are correct. Specifically, *E. coli* BL21 (DE3) wildtype cells were cultivated in M9 medium supplemented with 1 g/L yeast extract (M9 base medium). Different substrates such as ethylene glycol, glucose and ethanol were supplemented to assess the ability of wildtype *E. coli* cells to grow on each individual substrate. The concept underlying this approach is the comparison of the maximal optical density obtained during growth in M9 medium relative to that attained during growth in M9 supplemented with, for example, ethylene glycol. Growth experiments were conducted under aerobic conditions in 125 mL shake flasks with 25 mL growth medium. Results revealed that similar optical density was obtained for cells cultivated in M9 + 5 g/L ethylene glycol compared to M9 base medium (Figure A6a). This suggests that wildtype *E. coli* cells could not utilize ethylene glycol for growth. Examination of the substrate utilization profile revealed some consumption of ethylene glycol (i.e., 0.5 g/L over 47 hours), but this was not sufficient to support growth on the substrate (Figure A6b). Finally, glucose and ethanol served as positive controls, where results indicate that wildtype *E. coli* could grow on glucose and ethanol in M9 base medium.

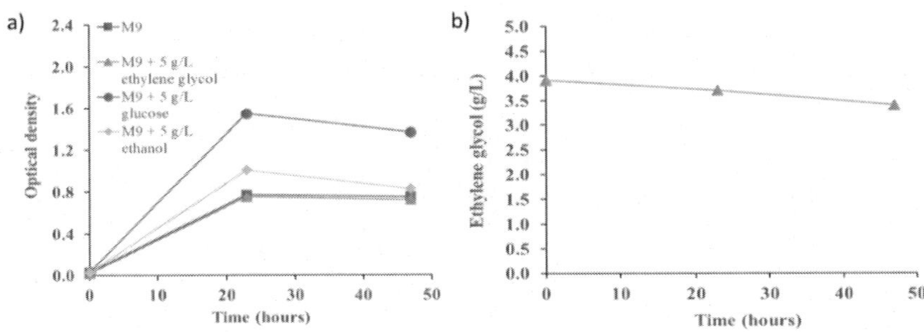

Figure A6. Assessing the ability of endogenous enzymes in wildtype *E. coli* in utilizing ethylene glycol. a) Similar growth profile with M9 base medium suggests that *E. coli* wildtype cells could not grow on ethylene glycol, b) Slight decrease in ethylene glycol concentration with time does not suggest significant uptake of the substrate.

Growing cells have been shown to deliver higher biotransformation efficiency in ethanol cofactor regeneration compared to resting cells. The effect comes about due to the coupling of the ethanol utilization pathway with central carbon metabolism through acetylCoA. Depletion of acetyl-CoA during cell growth would thus initiate the movement of flux from ethanol which leads to higher ethanol utilization and follow-on greater NADH regeneration. For ethylene glycol to have a similar benefit as ethanol, the terminal product of the heterologous ethylene glycol utilization pathway, glycolate, must integrate with central carbon metabolism. The logical connecting point for glycolate is glyoxylate. Two sets of enzymes are available in *E. coli* to achieve the conversion. One goes through the glycolate dehydrogenase complex (glcDEF), while the other achieve the same objective through glyoxylate reductase (ghrB). Based on the relative abundance of both sets of enzymes as profiled by RNA-seq analysis of transcriptome of *E. coli* (Table A3), glycolate dehydrogenase could not be the path for converting glycolate to glyoxylate due to low expression level of glcD, glcE and glcF. Alternatively, flux could flow from glycolate to glyoxylate due to appreciable expression of glyoxylate reductase (expression count: 51). Although lower than gyrase A, the expression level of glyoxylate reductase should be able to maintain some growth on ethylene glycol. However, limited growth and the ensuing reduced driving force would mean that cell growth may not yield as high a benefit on biotransformation efficiency as what was observed in growing cell ethanol cofactor regeneration powered butanone biotransformation.

Table A3. Expression level of different genes involved in glycolate to glyoxylate conversion in *Escherichia coli* wildtype cells

Gene	Function	Expression count
glcF	glycolate dehydrogenase, putative iron-sulfur	4
glcD	glycolate dehydrogenase, putative FAD-linked subunit	2
glcE	glycolate dehydrogenase, putative FAD-binding subunit	0
ghrB	glyoxylate reductase	51
gyrA	gyrase subunit A	574

6. Future Work

Similar to the ethanol utilization project, ethylene glycol utilization and its proposed use for regenerating NADH in support of a biotransformation reaction at the whole-cell level requires the delineating of a series of scientific questions that could lend the knowledge and insights that afford researchers an easy decision matrix for whether the NADH regeneration system is suitable for their systems as well as how to tune the system to maximize amount of NADH regenerated under particular conditions in order to improve biotransformation efficiency and product titer. Although most of the scientific questions are of an applied nature, their systematic elucidation through experimentation will nevertheless provide many insights into fundamental questions on physiology and differential gene expression in cells under varied conditions. Typically, an applied research project is held together by a series of questions revolving around a common theme. The approach differs from a fundamental basic science project that is driven by curiosity. However, effort will be expended to carefully think through the implications of findings from questions of an applied nature, particularly, whether their solution may yield new avenues for research or suggest a paradigm shift.

Preliminary work in this project has delineated the utility of HPLC as analytical tool of choice for determining the concentrations of various analytes pertinent to the biotransformation system. The next step taken was to understand if wildtype *E. coli* could grow on ethylene glycol as a way to assess if endogenous expression of fucO and aldA was low. A negative answer to the question validates the approach of over-expressing fucO and aldA on a plasmid in *E. coli* for conferring the capability to utilize ethylene glycol. This then set the stage for understanding the ability of *E. coli* to grow in aerobic and anaerobic environment with ethylene glycol as sole carbon source with over-expression of fucO and aldA on a plasmid. Subsequently, work would focus on demonstrating if the biotransformation system function as an integrated system in allowing ethylene glycol utilization to regenerate NADH in support of butanone biotransformation. Success in this step would usher in the next step in understanding possibility of toxicity effect from elevated concentrations of ethylene glycol and butanone in the biotransformation system. This would be followed by resting cell and growing cell biotransformation carried out in different growth medium to understand in details the conditions that yield the best biotransformation efficiency. Depending on how well ethylene glycol is utilized by the system, there may be opportunities for tuning the expression of fucO and aldA to improve substrate utilization rates and NADH regeneration. In addition, possibility of co-feeding of acetate and ethylene glycol in enhancing biotransformation efficiency, especially in growing cells would be explored. These steps in the research programme are outlined in a conceptual diagram below (Figure A7) and will be described in detail in the following paragraphs.

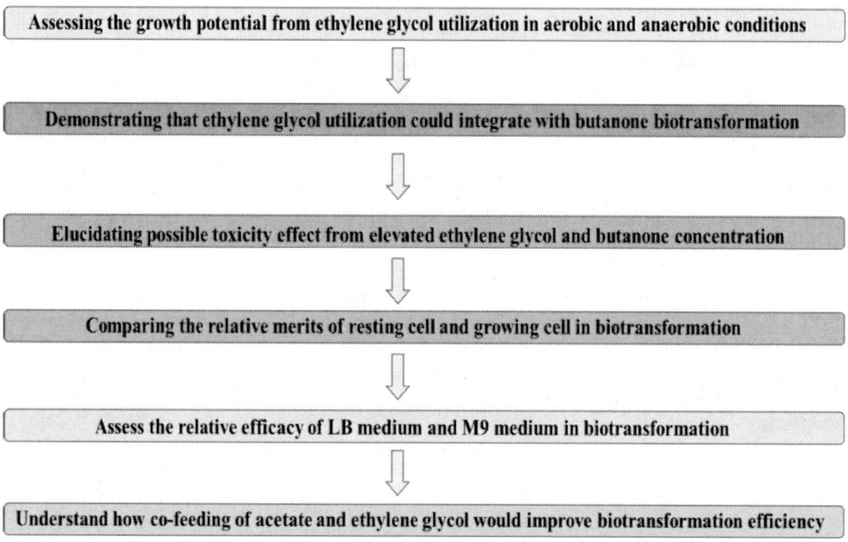

Figure A7. Conceptual flow of the ethylene glycol cofactor regeneration project.

Assessing the growth potential of ethylene glycol utilization in both aerobic and anaerobic conditions is important for gaining initial understanding of the kinetics of ethylene glycol utilization as well as how well glycolate is able to integrate with central carbon metabolism to support growth. The latter point is important given that the main route for channelling flux from glycolate into central carbon metabolism is through its conversion into glyoxylate. Better integration with central carbon metabolism as typified by greater flow of flux from glycolate or glyoxylate would thus mean better growth performance. Hence, assessing the growth performance of *E. coli* with ethylene glycol utilization pathway not only brings benefits in determining if the pathway is functional during growth in M9 medium with ethylene glycol as sole carbon source, it would also provide us with some initial bearing on how well the pathway has integrated with central carbon metabolism. Cell growth assay would be conducted at 37°C and 225 rpm in 125 mL shake flasks. Samples for optical density and HPLC analysis would be aliquoted at appropriate time-points (Estimated time required: 2 weeks).

Integration of the ethylene glycol utilization pathway and cpsADH enzyme in a single host would be the next logical step of the project to demonstrate the functionality of the system at the integrated level. To this end, co-transformation of the plasmid carrying ethylene glycol utilization pathway and cpsADH would be carried out in chemically competent *E. coli* BL21 (DE3) through the heat shock method. The strain obtained after biotransformation would be used in resting cell and growing cell biotransformation to assess the overall system level performance of the integrated biotransformation system. Cells would be precultivated in LB or M9 medium in resting cell biotransformation, or cultivated in LB or M9 medium in growing cell biotransformation. Both substrate (butanone) and co-substrate (ethylene glycol) concentrations would be 10 g/L. Glucose would serve as the positive control to check the overall functionality of the integrated biotransformation system with two plasmids. Resting cell biotransformation would run over a period of two days. Experiments will be conducted in 2.0 mL HPLC vials with 0.5 mL of reaction mixture. 89 mM phosphate buffer will serve as solvent for the biotransformation reaction (Estimated time required: 2 weeks).

Butanone has been demonstrated in the ethanol cofactor regeneration project to elicit a substrate toxicity effect in single-phase biotransformation. This aspect can be re-assessed again in the current project looking into ethylene glycol powered cofactor regeneration. Similar to the case in ethanol powered cofactor regeneration, there should be toxicity effect setting in at 20 g/L butanone with 10 g/L butanone as a suitable concentration for use in cofactor regeneration. On the other hand, the toxicity effect of ethylene glycol and its impact on biotransformation efficiency is relatively unknown. To help elucidate this aspect of the biotransformation system, a concentration range of ethylene glycol ranging from 5, 10 and 20 g/L ethylene glycol would be used for assessing toxicity effect associated with ethylene glycol. The key parameter of interest in this study is yield per unit OD, which is likely to decrease with onset of toxicity effect of ethylene glycol. Reduction in yield per unit OD would likely be perpetuated from decline in NADH regeneration, which should manifest effects on butanone uptake, which is the second parameter of interest in this study. Finally, toxicity effect of ethylene glycol may also negatively impact on its uptake and consumption, and this is the third parameter of interest (Estimated time required: 4 weeks).

Resting cell and growing cell biotransformations provide two perspectives of the problem where resting cell biotransformation holds more relevance for biocatalysis and growing cell biotransformation informs the utility of the ethylene glycol utilization pathway in NADH regeneration in metabolic engineering applications. This point is important as it is one of the future goals of this thesis to extend the ethylene glycol utilization pathway towards biomass and product formation. Resting cell biotransformation enjoys advantages in lower cost and ease of implementation and is more suitable for biocatalysis. But, it suffers from the inability of generating a growth-related driving force to aid in improving biotransformation efficiency. In particular, growing cell biotransformation could be tied in with other aspects of metabolic engineering where cell growth and running of the full gamut of metabolic pathways help yield the target product. From the process engineering and fundamental scientific perspectives, it would be interesting to compare the relative efficacy of both modes of biotransformation in enabling product formation, which provides direct evidence of whether ethylene glycol could operate better in resting cell or growing cell mode during biotransformation. Optimal concentrations of ethylene glycol together with butanone would be used as substrates. Cells for the resting cell biotransformation experiment would be concentrated two times, while those for the growing cell biotransformation would be diluted 10 fold (Estimated time required: 2 weeks).

Cell growth in different growth medium would likely elicit different gene expression pattern, which may have an impact on biotransformation efficiency. Experience from the ethanol utilization cofactor regeneration project indicates that M9 medium delivers higher biotransformation efficiency compared to LB medium, presumably due to the lower expression of NADH requiring enzymes that would divert regenerated NADH from biotransformation. LB and M9 medium would again be employed in this series of experiments given that they are well-researched representatives of rich and minimal medium, respectively. Both resting cell and growing cells would be employed to test the growth medium effect. In particular, growing cell mode place greater demand on the growth medium to support cell growth, which may have an impact on availability of cellular building blocks for synthesizing different enzymes (Estimated time required: 2 weeks).

From the theoretical perspective, ethylene glycol utilization could connect to glyoxylate through conversion from glycolate. But, glyoxylate remains a fringe molecule in central carbon metabolism without condensing with acetyl-CoA to form malate, an intermediate in tricarboxylic acid cycle. This reaction would not be productive given the small intracellular pool of acetyl-CoA, which will have major impacts on the driving force that could be generated from cell growth for biotransformation. One approach for providing a source of acetyl-CoA is through supplying acetate that could be converted into acetyl-CoA through acetyl-CoA synthetase (acs). Feasibility of this approach could be partially assessed through determining the relative abundance of acs as determined through RNA-seq analysis described above. Results indicate that there is low level expression of endogenous acs gene in *E. coli* (Table A4), which should be sufficient for driving formation of some acetyl-CoA that could be condensed with glyoxylate to form malate. Since the RNA-seq data is for *E. coli* K-12 at stationary phase, expression level for the acs gene at exponential phase may be higher, the co-feeding of acetate and ethylene glycol for improving biotransformation efficiency in growing cell is technically feasible. Resting cells, on the other hand, will serve as the negative control for the co-feeding of acetate and ethylene glycol to *E. coli* in aid of biotransformation (Estimated time required: 2 weeks).

Table A4. Expression level of endogenous acetyl-CoA synthetase gene in *E. coli* K-12 in minimal medium at stationary phase

Gene	Function	Expression count
acs	acetyl-CoA synthetase (AMP-forming)	33
gyrA	gyrase subunit A	574

Table A5. Timeline for the ethylene glycol NADH regeneration project

No.	Major tasks in the project	Time required
1	Assessing the growth potential from ethylene glycol utilization in aerobic and anaerobic conditions	2 weeks
2	Demonstrating that ethylene glycol utilization could integrate with butanone biotransformation	2 weeks
3	Elucidating possible toxicity effect from elevated ethylene glycol and butanone concentration	4 weeks
4	Comparing the relative merits of resting cell and growing cell in biotransformation	2 weeks
5	Assess the relative efficacy of LB medium and M9 medium in biotransformation	2 weeks
6	Understand how co-feeding of acetate and ethylene glycol would improve biotransformation efficiency	2 weeks
7	Comparison of biotransformation efficiency amongst ethylene glycol, glucose and ethanol utilization	2 weeks
8	Possible need to tune promoter strength of ethylene glycol utilization pathway	3 weeks
9	Plasmid construction	2 weeks
10	General troubleshooting	4 weeks
	Total:	25 weeks

Finally, depending on the substrate utilization efficiency for ethylene glycol during biotransformation, there may be a need to tune the strength of the promoter controlling expression of ethylene glycol utilization pathway genes. Higher expression level of fucO and aldA would improve ethylene glycol utilization and NADH regeneration. If higher transcriptional level is needed for the pathway genes, RNA-seq data can be consulted to identify suitable promoters. Specifically, genes with high expression level should have a corresponding promoter of high promoter strength. Given that most genes in *E. coli* are arranged in operon, care must be exercised in identifying the correct promoter for the gene. Experiments would also be conducted to assess the relative efficacy of the ethylene glycol utilization pathway in regenerating NADH compared to glucose dehydrogenase and ethanol utilization pathway. Overall, the total amount of time needed for the above proposed experiments is about 14 weeks. Plasmid construction would likely take another 2 weeks. Another 4 weeks would be factored in for troubleshooting problems encountered in the experimentation. Thus, about 20-25 weeks or 5-6 months would be needed to execute this project. A summary of the timeline is provided in Table A5.

CONCLUSION

While glucose dehydrogenase is the dominant NADH regeneration system in biocatalysis, there remains scientific value in exploring other NADH regeneration system mediated by the utilization of other substrates. One possible substrate, ethanol, has been explored in an earlier work in this thesis proposal. Another possibility comes by accident during the formulation of a two-step utilization pathway for ethylene glycol. Both steps of the pathway generate NADH from NAD^+, and theoretically could be used for cofactor regeneration. Formation of a metabolite, glycolate, that could integrate with central carbon metabolism through glyoxylate is another factor arguing for the use of the postulated ethylene glycol utilization pathway for cofactor regeneration since it does not end with a unnatural metabolite that in accumulating in the cell, would cause toxicity problem later on. Finally, exploring the NADH regeneration capability of the ethylene glycol utilization pathway would also illuminate the relative efficiency of pathway enzymes.

The ethylene glycol utilization pathway would be combined with cpsADH to form a plasmid-based system for butanone biotransformation in *E. coli* BL21 (DE3). This system enjoys the benefit of having all metabolites, substrates and products, easily detectable by a single HPLC run given that they are all water-soluble. Preliminary findings from the project have validated the analytical concept based on HPLC detection of ethylene glycol and glycolate. Although ethylene glycol's retention time of 13.9 mins is quite close to that of acetate (13.2 mins), separation of the two analytes should not be a problem. Given that both steps of the ethylene glycol utilization pathway are powered by genes native to *E. coli*, it is important to check the endogenous capability of *E. coli* in utilizing ethylene glycol. Results show that *E. coli* could not utilize ethylene glycol to support growth, a finding which is in sync with transcriptome analysis revealing that the expression level of endogenous fucO is low. Finally, transcriptome analysis also reveals that glycolate produced by ethylene glycol utilization could be connected to glyoxylate through endogenous glyoxylate reductase.

Planned experimental work sought to first demonstrate that a heterologous ethylene glycol utilization pathway could afford *E. coli* the ability to grow on the substrate, a fact which would also show that the pathway is functional. After which, attention would switch to performing initial integration work for both parts of the biotransformation system (i.e., cofactor regeneration and biocatalytic enzyme). Substrate toxicity effect, common in single phase biotransformation system, would be the next focus given uncertainty over the toxicity of ethylene glycol at elevated concentration. Resting cell and growing cell biotransformation, and comparison of their relative efficacy at biotransformation would be a major focus of the work. Next, growth medium effect that impact on gene expression pattern would be examined through work on rich medium (LB medium) and minimal medium (M9 medium). Finally, cofeeding of acetate and ethylene glycol would be explored to improve biotransformation efficiency in growing cells.

Overall, the proposed work should help us assess if ethylene glycol utilization could help regenerate NADH in support of a biotransformation reaction in *E. coli* as well as its relative efficiency compared to glucose dehydrogenase system and ethanol utilization pathway. From the applied perspective, the work should provide insights into the system parameters that govern the functioning of the biotransformation system such as appropriate substrate concentration and whether growing cell biotransformation would enable higher biotransformation efficiency compared to resting cell biotransformation.

Chapter 4

Future Project B: Teaching *Bacillus Subtilis* to Grow on the Unconventional Substrate Ethylene Glycol through Metabolic Engineering

1. Introduction

Central carbon metabolism of most microorganisms has been specifically tuned to utilize common substrates present in the environment. While some selective microbial species would have differentiated central carbon metabolic pathways, most of the microorganisms share high level of commonality in their core metabolic pathways. This led to the frequent use of common substrates such as glucose, xylose and glycerol in metabolic engineering as these substrates could be efficiently metabolized by endogenous enzymatic pathways to yield energy and carbon building blocks on which the biomass formation depends.

In recent years, focus in metabolic engineering has shifted to the exploration of unconventional substrates to first support biomass formation, followed by the derivation of useful chemicals and products from the carbon present in the substrate [61]. This emphasis on building a bioeconomy is driven in part by the desire to reduce society's dependence on fossil fuels for deriving many of its commodity chemicals and products. Work in this direction has generally focused on C1 compounds such as formate and carbon dioxide [62-69], the latter of particular interest given the need to find ways to utilize excess carbon dioxide in the atmosphere as mitigation strategy for climate change. However, there is less focus on C2 compounds such as ethanol and ethylene glycol.

Ethanol is more well-known as an end-product with role as a biofuel [70]. Its utility in NADH regeneration has been explored in the first project of this thesis. Focus in this segment of the work is on ethylene glycol. Although an obscure molecule, ethylene glycol could be derived from the hydrolysis of polyethylene terephthalate (PET) plastic or lignocellulosic waste. Hence, developing an utilization pathway for ethylene glycol is the first necessary step to explore the biotechnological potential of this C2 molecule. An utilization pathway has roles in plugging the carbon flux from ethylene glycol into central carbon metabolism for endogenous metabolism to help derive useful value-added products for the bioeconomy. For each given utilization pathway, much work remains in understanding how well the heterologous pathway integrates with endogenous metabolism as well as the extent in which the implemented pathway conducts crosstalks with endogenous enzymes that deplete precious carbon flux or lead to metabolic chokepoints. One role of metabolic engineering for substrate utilization is to identify these metabolic crosstalks, diagnose their bottlenecks and proffer a solution strategy to relieve metabolic chokepoints to allow as much substrate flux to flow to biomass formation or desired product.

This segment of the thesis work sought to implement ethylene glycol utilization in *Bacillus subtilis* as a proof-of-concept that the pathway is conceptually applicable to many microbial chassis in addition to *Escherichia coli*, where it was first implemented in Prof. Zhou's lab. Choice of *B. subtilis* arise from its status as Generally regarded as safe (GRAS), its naivety to ethylene glycol, and its use as a key biotechnological workhorse. The point on naivety to ethylene glycol is important as to the best of my knowledge, *B. subtilis* is unable to utilize the two-carbon substrate (from initial experimental results). Hence, implementing the ethylene glycol utilization pathway would be an important proof-of-concept that the pathway is conceptually generic, and is applicable to many biotechnological workhorses after some finetuning of promoters and ribosome binding sites for heterologous expression of the pathway genes.

As construed by my lab colleague, the ethylene glycol utilization pathway is a two-step pathway comprising fucO and aldA to sequentially process ethylene glycol to first glycoaldehyde and subsequently to glycolate. This original incarnation of the pathway has been successfully implemented in *E. coli* to support good cell growth with upwards of 20 g/L of ethylene glycol. Although the pathway ends at glycolate, endogenous glyoxylate reductase should be able to connect glycolate to glyoxylate, which help plugs the ethylene glycol utilization pathway to central carbon metabolism in *E. coli* (Figure B1).

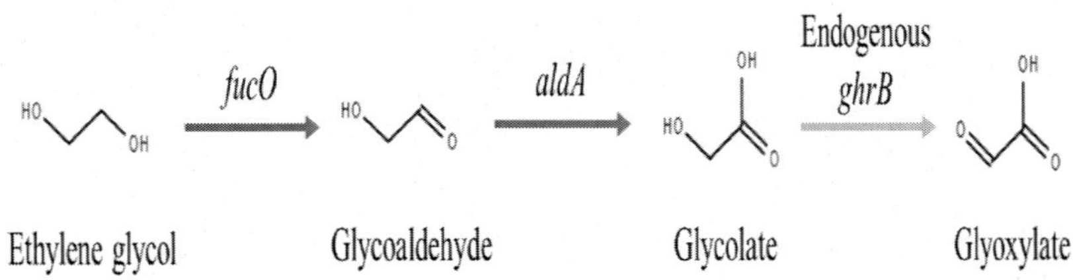

Figure B1. Architecture of the ethylene glycol utilization pathway as implemented in *E. coli*. Note the use of endogenous glyoxylate reductase to connect glycolate to central carbon metabolism metabolite, glyoxylate.

B. subtilis differs from *E. coli* substantially with respect for how the original ethylene glycol utilization pathway could connect to glyoxylate. In *B. subtilis*, glycolate could not connect with glyoxylate without heterologous expression of needed genes. More importantly, *B. subtilis* lacks the glycerate pathway to utilize the carbon in ethylene glycol. Hence, to afford the ability of achieving growth on ethylene glycol, substantial modifications and extension to the ethylene glycol utilization pathway is needed for implementation in *B. subtilis*. The microbial chassis chosen for this project is *B. subtilis* BS1A1, which is a derivative of *B. subtilis* 168.

Concept underlying the building of the ethylene glycol utilization pathway in *B. subtilis* is to channel flux from ethylene glycol beyond glycolate and into glyoxylate. This represents a reasonable path to enter central carbon metabolism. Glycolate dehydrogenase (glcDEF) from *Escherichia coli* would facilitate the movement of flux from glycolate into glyoxylate. But, at glyoxylate, the pathway meets a challenge in the relative lack of driving force for glyoxylate to condense with acetyl-CoA to form malate, a key intermediate in the tricarboxylic acid (TCA) cycle. This comes about possibly due to the lack of endogenous acetyl-CoA pool in *B. subtilis*, which forms the basis for the provision of a second substrate, ethanol to provide a carbon flux to augment the intracellular acetyl-CoA pool. Given the inability of endogenous enzymes in *B. subtilis* to convert ethanol to acetyl-CoA, the ethanol utilization pathway explored in the first topic of this thesis would be employed together with the ethylene glycol utilization pathway. Finally, glyoxylate and acetyl-CoA would be condensed by malate synthase A (aceB) from *E. coli* to form malate (Figure B2). Thus, the ethylene glycol utilization pathway in *B. subtilis* would be extended to three genes/enzymes, and will be encoded on an operon in a single plasmid. The ethanol utilization pathway, on the other hand, will be encoded on the second plasmid, and will have the same architecture as that used in the ethanol cofactor regeneration project.

Although linking ethylene glycol to malate through the metabolic branch point glyoxylate has effectively connected ethylene glycol utilization to central carbon metabolism, possibility also exists of connecting ethylene glycol to other branches of metabolism. This arises because of the role that glyoxylate plays as a metabolic branch point. Specifically, flux from glyoxylate could be diverted to glycine, malate and 2-phosphoglycerate. The last of which is the focus of the second part of the project, particularly for understanding the technical feasibility of connecting ethylene glycol to 2-phosphoglycerate and glycolysis. Such an endeavour would also provide growth data that would afford a direct comparison of the relative utility of connecting ethylene glycol to tricarboxylic acid cycle (i.e., malate intermediate) or glycolysis (i.e., 2-phosphoglycerate intermediate). Variables that could impact on biomass formation efficiency include how well carbon flux from ethylene glycol could be channelled to biomass precursors, as well as the existence of metabolic branch point that divert carbon flux away from biomass formation.

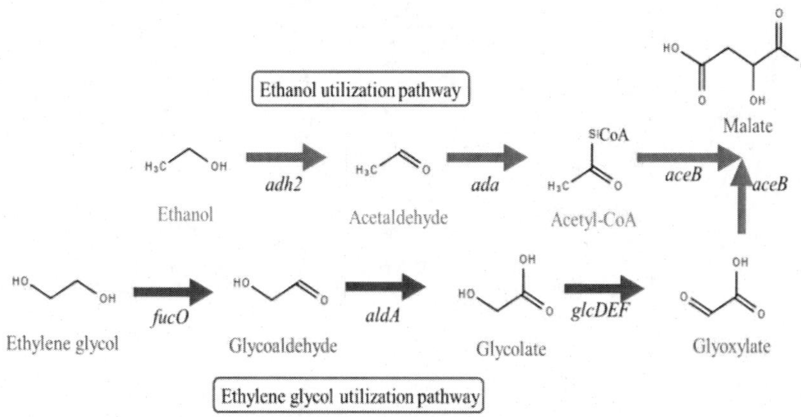

Figure B2. Architecture of the ethylene glycol utilization scheme in *B. subtilis* which require a pool of acetyl-CoA provided by ethanol for allowing glyoxylate to condense into malate and enter the tricarboxylic acid cycle. Acetyl-CoA pool could also be provided by acetate through endogenous acetyl-CoA synthetase in *B. subtilis*.

The metabolic route from ethylene glycol to 2-phosphoglycerate include 6 steps and enzymes, and comprise the pathway linking ethylene glycol to glyoxylate. All pathway genes selected are endogenous to *E. coli*. At glyoxylate, the pathway follows the route to glycolysis rather than tricarboxylic acid cycle. Glyoxylate is converted to tartronate semialdehyde through glyoxylate carboligase (gcl). This is followed by conversion of tartronate semialdehyde to glycerate through tartronate semialdehyde reductase (glxR). Finally, glycerate is phosphorylated to form 2-phosphoglycerate by glycerate-2-kinase (glxK). As designed, the pathway linking ethylene glycol to 2-phosphoglycerate is lengthy, and opportunities exist for breaking it up into two modules, which can be encoded in two plasmids. The suitable branch point is again glyoxylate. By separating a long pathway into 2 independent modules, possibilities exist of doing independent metabolic tuning of the two pathway segment, which would help reduce experimental effort compared to tuning expression of individual genes.

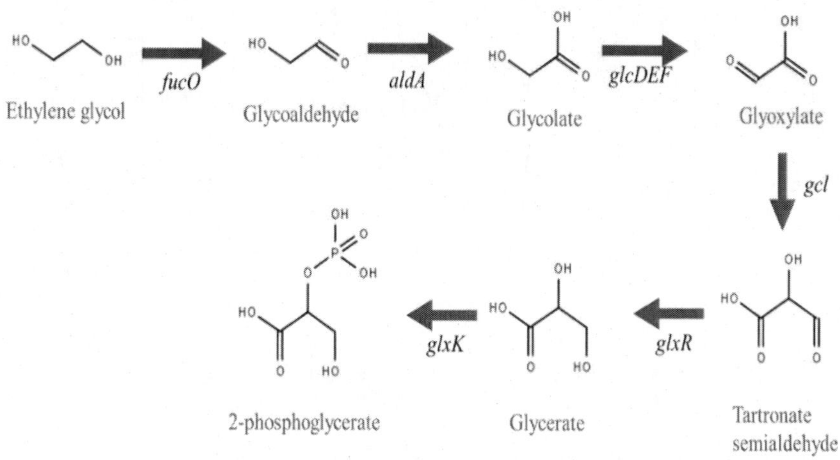

Figure B3. Architecture of utilization pathway linking ethylene glycol to 2-phosphoglycerate through the metabolic branch point glyoxylate, whose flux could also be diverted to other metabolites. This represents a route linking ethylene glycol to glycolysis.

Overall, teaching *B. subtilis* to utilize the unconventional substrate ethylene glycol through metabolic engineering would provide a platform on which first biomass would be produced from the substrate, and subsequently the derivation of useful products from the substrate. Ethylene glycol need not be an expensive source of substrate as it is the hydrolysis product of PET plastic waste as well as that of lignocellulosic waste. What matters in engineering a substrate utilization pathway for ethylene glycol is the selection of a metabolic route that could expeditiously (in as few steps as possible) convert the substrate into biomass without significant loss of carbon. Ethylene glycol utilization as facilitated by extant enzymes in *E. coli* would link the substrate to glyoxylate. As an important metabolic branchpoint, many routes can emanate from the 2-carbon molecule, thereby, offering many options for connecting ethylene glycol flux to different branches of central carbon metabolism. The initial choices made in this project proposal are malate (TCA cycle) and 2-phosphoglycerate (glycolysis). By comparing the relative growth potential of the two different metabolic routes that originate from ethylene glycol, this project should inform on the relative efficacy of the two routes in generating biomass and activating production of different biomass precursors. Such information is crucial for understanding the better performing route for biomass formation, but, would also help calibrate our theoretical understanding of the two routes' relative performance.

2. THEORETICAL UNDERPINNINGS OF THE PROJECT

Understanding the redox balance of a pathway is critical for success in metabolic engineering. Ideally, redox balance between reducing and oxidizing equivalent of a cofactor would help maintain the redox environment in the cell, thereby, enabling better cell growth. To this end, the pathways involved in the co-feeding of ethanol and ethylene glycol for facilitating ethylene glycol utilization was examined (Figure B4). In terms of redox cofactors, the ethylene glycol utilization pathway has two NADH production steps at fucO and aldA. Similarly, the ethanol utilization pathway also has two NADH production steps at adh2 and ada. Altogether, the co-feeding of ethanol and ethylene glycol would produce a total of 4 NADH per turn of both cycles. These represents a significant source of reducing power which can be tapped to power formation of cellular building blocks, and biomass precursors. Growth is necessary to provide a sink for the excess NADH generated by both pathways. In this case, growth medium may play an outsized role in determining the efficiency in which ethylene glycol and ethanol are utilized. Specifically, a growth medium with sufficient nutrients to support good growth would provide a larger sink for consuming the excess NADH produced, which would translate into higher substrate utilization and better growth.

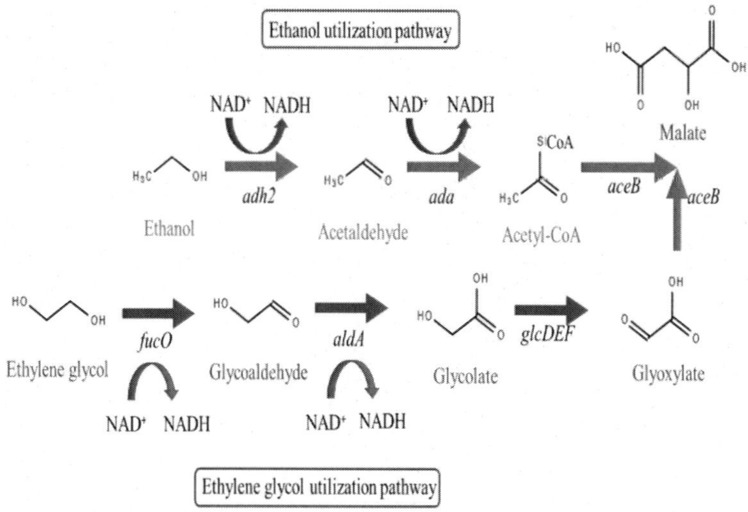

Figure B4. Analysis of the redox balance of the ethylene glycol and ethanol utilization pathways reveals a preponderance of NADH production steps, which are not balanced by downstream enzymatic steps. However, as cell growth and biomass formation are the ultimate aims of this project, these NADH requiring processes should be able to consume the excess NADH generated by the pathways, and help ensure redox balance.

The above scheme connects ethylene glycol to malate in TCA cycle through condensing one ethanol-derived acetyl-CoA with one ethylene glycol-derived glyoxylate. For the metabolic route to work, there must be an efficient enzyme to convert malate into oxaloacetate in order for TCA cycle to function properly. Malate dehydrogenase, an enzyme candidate that fulfills the enzymatic task, thus come into the picture. To test whether the construed utilization pathway would work, growth experiments need to be conducted with malate as sole carbon source in M9 minimal medium to help assess if *B. subtilis* wildtype cells could efficiently convert malate into biomass, and thus, provide support that the envisioned pathway would work.

Similar cofactor dependence analysis is also needed for the pathway that connects ethylene glycol to 2-phosphoglycerate. In terms of NADH, fucO and aldA each generates one NADH prior to glyoxylate branch point (Figure B5). Downstream of glyoxylate, 1 NADH is generated by tartronate semialdehyde reductase. The pathway is also a net energy consuming pathway with the need for 1 ATP molecule to power the phosphorylation of glycerate with glycerate-2-kinase. Overall, the metabolic route does not have any NADH consumption step, thereby, leaving an imbalance of redox cofactor in NAD^+/NADH. Such imbalance can be corrected with the depletion of NADH by growth processes provided the imbalance is not too severe. At net production of 3 NADH per turn of the pathway, the imbalance is not too severe, but only actual experiments and troubleshooting will reveal the answer. ATP demand from the pathway is also not high. Hence, the pathway should have a reasonable chance to drive flux from ethylene glycol to 2-phosphoglycerate. What is more of a concern is the kinetics of the enzymes in the pathway and the overall productivity. Such influences on the pathway would become evident with experimental characterization revealing overall system performance.

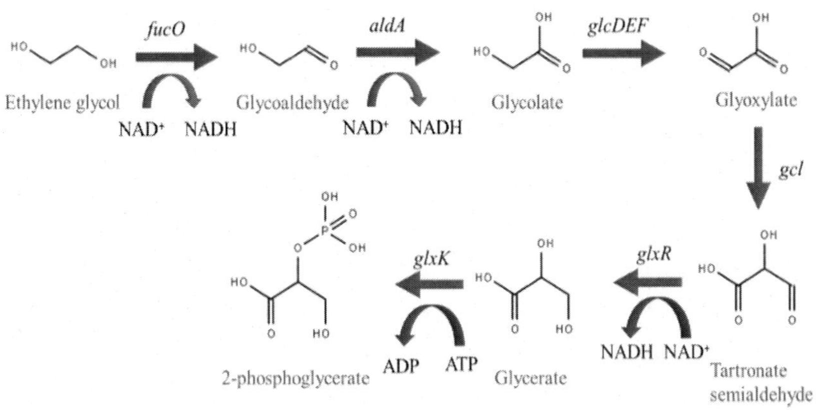

Figure B5. Analysis of cofactor balance of the metabolic route linking ethylene glycol to 2phosphoglycerate through glyoxylate. Overall, 3 NADH is generated with the consumption of 1 ATP.

Endogenous genes in *B. subtilis* cells may be available to perform some of the reactions in the metabolic route linking ethylene glycol and malate. To examine this question, the gene database and RNA-seq transcriptome of *Bacillus subtilis* 168 cultivated in LB medium was used to obtain information about the genetic repertoire of the bacterium as well as the gene expression pattern in LB medium. Given that use of a rich medium should help activate the expression of more genes compared to a minimal medium, the gene expression pattern available in cells grown in LB medium should provide an upper estimate of the genetic capability of the organism. In this endeavour, particular emphasis is placed on whether there is an endogenous glycolate oxidase in *B. subtilis* for converting glycolate to glyoxylate as well as the expression level of malate synthase that help condense acetyl-CoA with glyoxylate to form malate that enters the TCA cycle. Table B1 shows the expression level of enzymes in *B. subtilis* 168 that could help bring metabolic flux from glycolate to malate with acetyl-CoA provided by ethanol utilization. The candidate enzymes are equivalent forms of glycolate dehydrogenase in *E. coli* and malate synthase A. In the case of *E. coli* glycolate dehydrogenase, endogenous glycolate oxidase may fit the bill, but its expression level is zero both for both subunits. Such an observation explains why glyoxylate shunt is not operable in *B. subtilis* [71]. Hence, there is a need to express *E. coli* glycolate dehydrogenase in *B. subtilis* to reconstitute the pathway. On the other hand, an equivalent form of malate synthase A in *E. coli* could not be found in *B. subtilis*. This is not impossible as glyoxylate shunt has been reported to be not operable in *B. subtilis*, and examination of genetic repertoire of *B. subtilis* suggested the absence of malate synthase that condense acetyl-CoA with glyoxylate. Hence, there is a need to express malate synthase A from *E. coli* in *B. subtilis* to achieve the channelling of flux from ethylene glycol to a central carbon metabolite in the Gram-positive bacterium.

Table B1. Expression level of enzymes in *B. subtilis* 168 targeted for expression in the metabolic route linking ethylene glycol to malate

Gene	Enzyme	*Bacillus subtilis*	Expression count
glcF	glycolate oxidase (iron-sulfur subunit)	Present	0
glcD	glycolate oxidase subunit	Present	0
	malate synthase	Not present	

Similar analysis was conducted to assess the need for heterologous expression of the target enzymes involved in converting ethylene glycol to 2-phosphoglycerate in *B. subtilis* (Table B2). Specifically, glycolate carboligase (*gcl*) from *E. coli* is not present in the *B. subtilis* genome, and would need to be heterologously expressed. The same also applies to tartronate semialdehyde reductase from *E. coli*. Its absence in *B. subtilis* meant that it has to be heterologously expressed. Finally, analysis of the gene database found a D-glycerate kinase (glxK) which may serve the same function as glycerate-2-kinase in *E. coli*. However, readout from the RNA-seq transcriptome of *B. subtilis* cultivated in LB medium showed that it was not expressed, which suggests that the components of the glycerate pathway are either missing or not functional in *B. subtilis*. Hence, there is also a need to express glycerate-2-kinase from *E. coli* in *B. subtilis*. All three genes of the glycerate pathway that links glyoxylate to 2phosphoglycerate would be expressed as an operon on a plasmid. This constitutes as a gene expression module in pathway construction, and affords an easy way to modulate the expression level of all pathway genes through tuning the promoter controlling the operon's expression level.

Table B2. Presence/absence and expression level of enzymes in *B. subtilis* 168 targeted for expression in the metabolic route linking ethylene glycol to 2-phosphoglycerate

Gene	Enzyme	*Bacillus subtilis*	Expression count
gcl	glycolate carboligase	Not present	
glxR	tartronate semialdehyde reductase	Not present	
glxK	D-glycerate kinase	Present	0

Given the length of the pathway linking ethylene glycol to either malate or 2phosphoglycerate, need exists in separating the long pathway into different modules. For the ethylene glycol to malate route, the choice is easy, the pathway from ethylene glycol to glyoxylate would be encoded on one plasmid. Ethanol utilization pathway together with malate synthase A would constitute another pathway module encoded on another plasmid. On the other hand, the pathway linking ethylene glycol to 2-

phosphoglycerate could be divided into 2 modules at the metabolic branch point, glyoxylate. Pathway leading from ethylene glycol to glyoxylate would constitute one module, while the metabolic route leading out from glyoxylate to 2-phosphoglycerate would be another module. Doing it this way affords ready tuning of expression level of genes in the two separate branches of the entire pathway.

Evidence for the ability of *B. subtilis* to utilize ethylene glycol for growth would naturally comes from observations of growth during cultivation of *B. subtilis* in M9 minimal medium supplemented with ethylene glycol. But, in order to account for the carbon flow in the metabolic network of the bacterium as well as gaining an understanding of how well the carbon flux flow from ethylene glycol towards central carbon metabolism and the route taken. One approach to verify that the metabolic flux did indeed flow in the predicted manner as construed in our pathway engineering comes from a natural experiment in *B. subtilis*. Specifically, *B. subtilis* could not synthesize glyoxylate from glycolate through endogenous metabolism. Similarly, the bacterium is also unable to condense acetyl-CoA with glyoxylate to form malate. Hence, if *B. subtilis* could grow on ethylene glycol after heterologous expression of genes that construct an enzymatic pathway linking ethylene glycol to malate, it can be said with reasonable confidence that metabolic flux did flow along the constructed pathway to enable cell growth. The same argument applies to the pathway linking ethylene glycol to 2phosphoglycerate. Given the lack of enzymes enabling the progressive stepwise conversion of glyoxylate to 2-phosphoglycerate, ability of *B. subtilis* to grow on ethylene glycol in a strain with the above pathway would mean that metabolic flux has flowed along the constructed pathway to allow carbon flux from ethylene glycol to connect to glycolysis at the entry point of 2-phosphoglycerate. From there, carbon flux would flow up and down glycolysis to power other branches of central carbon metabolism to produce biomass precursors for cell growth. Isotope labelling on the carbon atom in the substrate is commonly used in metabolic engineering to (i) investigate the flow of carbon in a constructed pathway, (ii) to check for unanticipated diversion of flux, and (iii) to confirm that cell growth was predominantly powered by carbon flux from the substrate. The approach could be utilized in this study to check for amount of flux that "leaks" from the constructed pathway, and whether biomass formation was due to carbon flux channelled by the constructed metabolic pathway.

3. LITERATURE REVIEW

Ethylene glycol is a potential substrate for biotechnology as it could be converted into glycolate and glyoxylate to enter central carbon metabolism. Source of PET could be from hydrolysis of lignocellulosic waste or polyethylene terephthalate (PET) waste [72]. A check on the academic literature reveals that most work revolving around ethylene glycol uses the polymerized form instead of the monomer. But, reports do exist on the fermentation of ethylene glycol by *Clostridium glycolicum*, which requires 15 amino acids and the vitamins biotin and pantothenate for optimal growth on the substrate [73]. A study details the anaerobic degradation of ethylene glycol and polyethylene glycol by methanogenic consortia obtained from sewage sludge [74]. Specifically, a mixture of microorganisms could degrade ethylene glycol monomer units in polyethylene glycol to monomers and subsequently to ethanol and subsequently to acetate [74]. Another study documented the anaerobic biodegradation of ethylene glycol by microbes in the rhizosphere of poplar and willow trees with complete conversion of ethylene glycol using nitrate and sulphate as terminal electron acceptors [75]. Psychotropic bacterium such as *Pseudomonas fluorescens* has also been reported to degrade ethylene glycol with the substrate inducing the expression of ethylene glycol shock and acclimation proteins [76]. Besides degradation of ethylene glycol, there has also been studies on the biosynthesis of the molecule [77]. Specifically, by channelling carbon flux from xylose through glycoaldehyde, ethylene glycol was synthesized at appreciable titer of 11.7 g/L for 40.0 g/L of xylose feed in *Escherichia coli* [78]. Other organisms such as *Corynebacterium glutamicum* and yeast have also been used in synthesizing ethylene glycol from sugars [77]. Overall, relatively few studies have looked into the biotechnological potential of ethylene glycol. To the best of my knowledge, there are no reports of utilization of ethylene glycol by *B. subtilis*.

Glycolic acid or glycolate is one of the intermediates in the proposed ethylene glycol utilization pathway. It is a valuable chemical as it has found use in the dyeing and tanning industries as well as the food processing industry. Work has been done using resting cells of *Gluconobacter oxydans* DSM 2003 for conversion of ethylene glycol to glycolic acid. Conversion rate is 86.7% for 48 hrs [79]. The approach demonstrated the thermodynamic and biological feasibility of the ethylene glycol to glycolic acid route. *E. coli, Corynebacterium glutamicum* and yeast have also been explored as production host for glycolic acid with sugars as source of carbon flux [77]. Hence, information from the literature has indicated that they may be commercial merits in developing glycolate as a product of ethylene glycol utilization via a two-gene pathway. This is certainly possible in the context of *B. subtilis*, which lacks enzymes for converting glycolate to glyoxylate that enters the tricarboxylic acid cycle. But, more broadly, terminating the pathway from ethylene glycol to glycolate could not connect the carbon flux to central carbon metabolism, which is necessary for deriving the multitude of biomass precursors important for powering growth. Thus, focus of this project remains the derivation of pathways for connecting ethylene glycol to metabolites in central carbon metabolism.

As mentioned above, ethylene glycol could be degraded under anaerobic conditions or low temperatures by bacteria. What is of interest in this project is the utilization of ethylene glycol. To this end, one interesting route different from the one proposed here is the degradation of ethylene glycol to acetaldehyde and acetate by *Mycobacterium* E44 [80]. This route offers a path to acetyl-CoA through acetate, and may be considered in later part of this project if time permits. The route from ethylene glycol to glycolate was first discovered in *E. coli* when mutants were discovered to grow on propylene glycol [8]. These mutants could also grow on ethylene glycol by first converting it to glycoaldehyde, which was subsequently converted to glycolate that enters central carbon metabolism.

This pathway could be extended to malate and glycerate in *E. coli* using endogenous enzymes, and provides inspiration for the construction of the ethylene glycol to malate route in *B. subtilis*. More recently, expression of glyoxylate carboligase (gcl) and tartronate semialdehyde reductase (glxR) conferred a growth phenotype to *Pseudomonas putida* exposed to ethylene glycol. Further expression of glycolate oxidase removed a metabolic bottleneck, and enabled consumption of high concentration of ethylene glycol [81]. Such a pathway linking ethylene glycol to glycerate is similar to the one proposed here for converting ethylene glycol to 2-phosphoglycerate.

Implementation of the ethylene glycol utilization pathway is only the first part of the story in enabling the substrate to serve as sole carbon and energy source. To achieve the second part of the endeavour, ethylene glycol utilization must connect to endogenous central carbon metabolism to afford distribution of carbon flux from ethylene glycol to different biomass precursors in the cell metabolic network to enable cell growth. To do this, the microbial chassis chosen should not have significant deficiencies in metabolic network, particularly enzymes involved in central carbon metabolism. Experimental studies have illuminated that *B. subtilis* may have a deficiency in synthesizing succinate and citrate, which forces their inclusion in a minimal chemically defined medium for the organism (data and analysis in later sections of this report). This necessarily define a need for conducting a review of other possible metabolic deficiency in this otherwise good biotechnology workhorse useful for secreting many recombinant proteins. A search with the academic literature reveals no reports of metabolic deficiency in *B. subtilis*. However, this does not mean that *B. subtilis* has the full metabolic repertoire for utilizing a variety of substrates. Specifically, a study has indicated that *B. subtilis* does not have the glyoxylate shunt; thereby, requiring engineering effort to allow it to grow on overflow metabolite [71]. Despite the lack of glyoxylate shunt and glycerate pathway, *B. subtilis* remains a suitable chassis for building and connecting an extended ethylene glycol utilization pathway for testing the hypothesis of whether carbon entry point in central carbon metabolism play an important role in determining growth efficiency.

4. Materials and Methods

4.1. Bacterial Strains Used

Bacillus subtilis BS1A1 wildtype strain in the lab was chosen for this work.

4.2. Chemicals and Growth Medium Used

Ethanol, ethylene glycol and glucose were purchased from Sigma-Aldrich and used without further purification. Composition of LB Miller medium is (g/L): Tryptone, 10.0; Yeast extract, 5.0; NaCl, 10.0. Composition of M9 medium with 1 g/L yeast extract is (g/L): K_2HPO_4, 6.8; KH_2PO_4, 3.0; NH_4Cl, 1.0; NaCl; 2.0; Yeast extract, 1.0.

4.3. Growth Assessment Test for Ethylene Glycol as Growth Substrate

B. subtilis BS1A1 wildtype strain was chosen for assessing whether endogenous enzymes in *B. subtilis* could facilitate the utilization of ethylene glycol. Glycerol stock culture of the wildtype *B. subtilis* strain served as inoculum for a seed culture comprising 10 mL of LB medium in a 125 mL glass shake flask. The seed culture was cultivated at 30°C and 225 rpm for 16 hrs, after which 0.25 mL of seed culture served as inoculum for 25 mL of M9 + 1 g/L yeast extract medium (M9 base medium) in a 125 mL glass shake flask. Similarly, the seed culture also served as inoculum for M9 base medium supplemented with different carbon sources such as

5 g/L ethylene glycol, 5 g/L ethanol, and 5 g/L glucose. The experiment cultures were cultivated at 30°C and 225 rpm. The incubator used was N-Biotek NB-205 Incubator-shaker.

At appropriate time points, aliquots were withdrawn from the experiment culture for both optical density and HPLC analysis. Optical density was measured with an UV-visible spectrophotometer at 600 nm (Ambersham Biosciences). Samples were diluted with deionized water if the absorbance exceeded 1. For HPLC analysis, 0.5 mL of sample was centrifuged at 20000xg for 2 mins, 200 μL of the supernatant was carefully pipetted out and filtered through a 0.22 μm nylon filter into a 300 μL plastic HPLC vial for analysis.

4.4. HPLC Analysis Conditions

Ethanol, glucose and ethylene glycol were the target metabolites for HPLC analysis. High performance liquid chromatography was conducted with an Agilent 1260 series instrument equipped with a Bio-rad Aminex HPX-87H column. 5 μL of sample was injected into the column and eluted with 5 mM sulphuric acid at 0.7 mL/min. Isocratic elution was used and the column was maintained at 50°C. Analytes were detected by a refractive index detector.

4.5. Constructing a Gene Database Comprising All Genes in *B. Subtilis* 168

B. subtilis 168 is taken as a reference strain for the species. In this part of the analysis, its annotated genome sequence was downloaded from Genbank (NCBI Reference Sequence: NC_000964.3). The annotated genome sequence was parsed using MATLAB, and relevant information such as gene name, gene function, and gene sequence was used in constructing a gene database for the organism.

4.6. RNA-Seq Data Analysis

RNA-seq transcriptome analysis was called upon to assess the plausibility of certain reactions in *B. subtilis* through providing the relative abundance of different genes in the genome of the bacterium. To this end, the RNA-seq dataset of *B. subtilis* 168 cultivated in LB medium (Array Express accession number E-GEOD-74926) was downloaded from Array Express and processed by an in-house MATLAB RNA-seq transcriptome processing software. Briefly, the approach entails the reading of a sequenced read, and aligning with all genes in the genome of *B. subtilis* 168. If the sequenced read aligns with a gene, the expression count of the gene would be increased by one. In total, 0.4 million reads were processed to yield the final transcriptional picture of *B. subtilis* 168 in LB medium.

4.7. Constructing an Expression Vector for Heterologous Gene Expression in *B. Subtilis*

Unique genetic engineering tools exist for the model Gram-positive bacterium, *B. subtilis*. Specifically, promoters from *E. coli* generally do not work in *B. subtilis*. Hence, work started to modify a plasmid vector useful for expressing ethanol utilisation pathway genes in *E. coli* for use in *B. subtilis*. Specifically, the pthrC autoinducible promoter was replaced with PgyrA from *B. subtilis*. PgyrA was chosen as it has relatively high promoter strength as gauged by the expression level of gyrase A gene in the transcriptome of *B. subtilis*. Annotated genome sequence of *B. subtilis* was used in identifying promoter region of the gyrA in the bacterium. Using appropriate primers, a 600 bp region of the genome region upstream of gyrA was cloned through colony PCR in the thermocycler (Biorad, T-100). Specifically, 1 μL of a 16 hour *B. subtilis* culture in LB Miller medium was used as template for PCR amplification. PCR cycling conditions were: 98°C for 5 mins, followed by 34 cycles of (98°C for 8 sec, 55°C for 15 sec, and 72°C for 60 sec), with final extension of 72°C for 5 mins. The PCR product was clean-up through gel electrophoresis (130V for 30 mins, Biorad power pack) followed by gel purification using Thermo Fisher Scientific gel purification kit (GeneJet Gel Extraction Kit, Cat No: #0692). The resulting promoter fragment was ligated with the plasmid backbone using a ligase-free ligation method developed in the lab that includes incubation with 10 mM $MgCl_2$ in a thermocycling step in a PCR thermocycler (Biorad, T-100). The cycling programme was as follows: 80°C for 1 min, 68°C for 10 mins, and infinite hold at 4°C. Plasmid construction method was developed in the lab, and is known as the GT-standard [37]. Primers used in the experiment are as described in Supplementary Table A1 below.

Supplementary Table A1. Primers used in cloning PgyrA from *Bacillus subtilis*

Promoter	Primer Name	Sequence
PgyrA	PgyrA-Forward	5'-agaacactgctgttaacgttcttttaca-3'
PgyrA	PgyrA-Reverse	5'-ttaaaaaacctccctgaagtatcacatgaa-3'

4.8. Transformation of Ligated Plasmid into *B. Subtilis* through Induction of Natural Competence

A stationary phase culture of *B. subtilis* (18 to 24 hrs of incubation time) after inoculation of glycerol stock culture in 10 mL LB Miller medium with incubation conditions at 30°C and 225 rpm served as source of competent cells. The workflow starts with aliquot of 1.6 mL of culture in a 2.0 mL microcentrifuge tube. The contents of the microcentrifuge tube was centrifuged at 3300 x g for 2 minutes in a benchtop centrifuge (Eppendorf, Germany). The supernatant was discarded. 1 µL of purified plasmid was added to the cell suspension at the bottom of the microcentrifuge tube with flicking of the tube for mixing cells and plasmid. 0.9 mL of SOC medium and 100 µL of 20% xylose solution was subsequently added to the cell suspension. The cell suspension was vortexed, and the cells incubated at 30°C and 225 rpm for 2 hours. After 2 hour of incubation, 200 µL of cell suspension was aliquoted for plating on

LB agar plate with appropriate antibiotics. Agar plates were incubated at 30°C overnight for the cells which have taken up the plasmid to form colonies.

5. Preliminary Findings

Initial experiments in the project sought to assess if *B. subtilis* BS1A1 could utilize ethylene glycol as well as other alternative substrate such as ethanol. This is in preparation for the co-feeding strategy where both ethanol and ethylene glycol would supply carbon for building malate that enters the tricarboxylic acid (TCA) cycle to power central carbon metabolism. Another facet of the experiment is its role in checking if the postulations derived from transcriptome and gene repertoire analysis is correct. Specifically, wildtype *B. subtilis* should not be able to uptake and utilize ethylene glycol. On the other hand, *B. subtilis* should be able to uptake ethanol, but could not grow on the substrate.

5.1. Initial Growth Experiments to Assess Substrate Scope of *B. Subtilis* Wildtype Cells

In aerobic growth experiments carried out in M9 medium (with 1 g/L yeast extract) at 30°C and 225 rpm assessing the substrate scope of *B. subtilis,* the bacterium could not grow on ethylene glycol and ethanol (Figure B6a). In fact, there was no utilization of ethylene glycol over a period of 3 days, which suggests that *B. subtilis* does not possess enzymes able to convert ethylene glycol (Figure B6b). Ethanol, on the other hand, could be uptake into the cell (data not shown), but could not support additional growth beyond that provided by nutrients in the M9 base medium. This indicates that the ethanol catabolism pathway did not yield acetyl-CoA such as the case in endogenous adhE of *Escherichia coli*, nor did the pathway connect to any growth promoting intermediates in central carbon metabolism. Hence, opportunities open up for the engineering of a growth promoting ethanol utilization pathway in *B. subtilis*. Bioinformatic search for the enzymes that participated in ethanol catabolism in *B. subtilis* would be an interesting angle for the research.

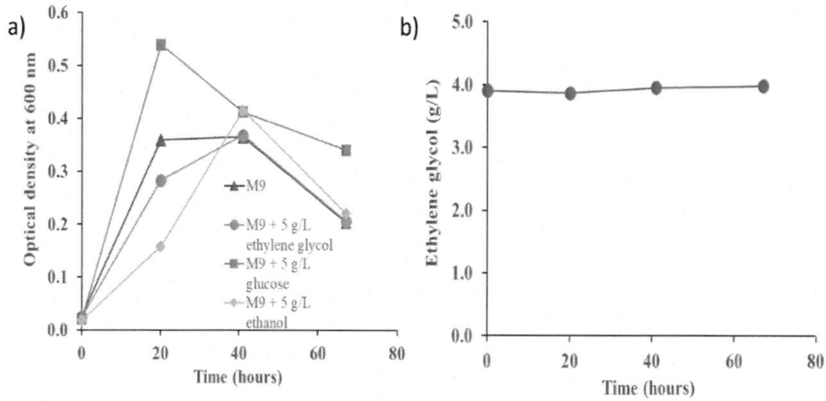

Figure B6. Aerobic growth assays designed to assess the substrate scope of *B. subtilis* wildtype cells in M9 medium with 1 g/L yeast extract at 30°C and 225 rpm. a) Evolution of optical density for cells grown in M9 medium supplemented with different carbon substrates, b) Substrate utilization profile of ethylene glycol.

Finally, growth experiments in M9 base medium supplemented with alternative substrates suggest that *B. subtilis* may have a metabolic defect in amino acid or vitamin synthesis that hampers its growth in minimal salt medium. Specifically, relatively little additional growth above the baseline was observed in M9 supplemented with 5 g/L glucose which serve as the positive control. This is in stark contrast to the case for *Escherichia coli* in the same medium under aerobic conditions (Figure B7a). Lack of growth with glucose could only suggests that *B. subtilis* need additional sources of vitamins and amino acids beyond what is available in 1 g/L yeast extract. More importantly, there is a deficiency in uptake of glucose by *B. subtilis* which may be due to lack of expression of genes involved in glucose uptake in minimal medium (Figure B7b). Hence, further experiments could be conducted in rich LB medium or M9 medium supplemented with complete supplement mixture (CSM).

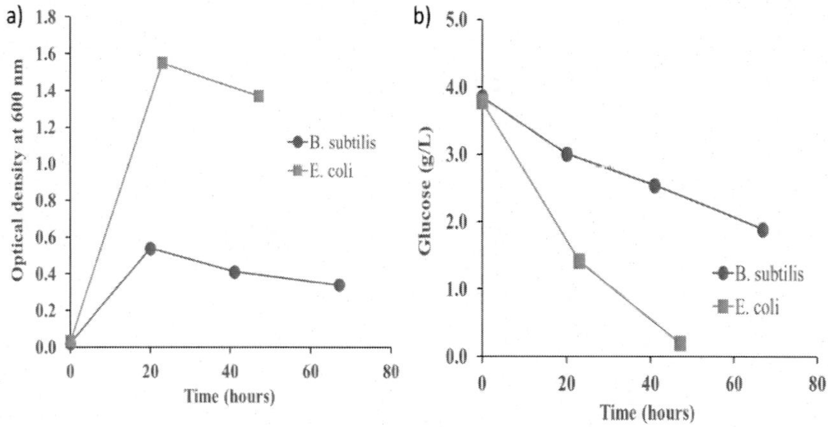

Figure B7. Comparison of growth performance of *B. subtilis* and *E. coli* in M9 medium with 5 g/L glucose. *B. subtilis* was cultivated at 30°C while *E. coli* was grown at 37°C. a) *B. subtilis* exhibited a growth defect during growth in M9 glucose medium compared to *E. coli*, b) Growth defect in *B. subtilis* in M9 glucose medium likely arise due to its deficiency in uptake of glucose.

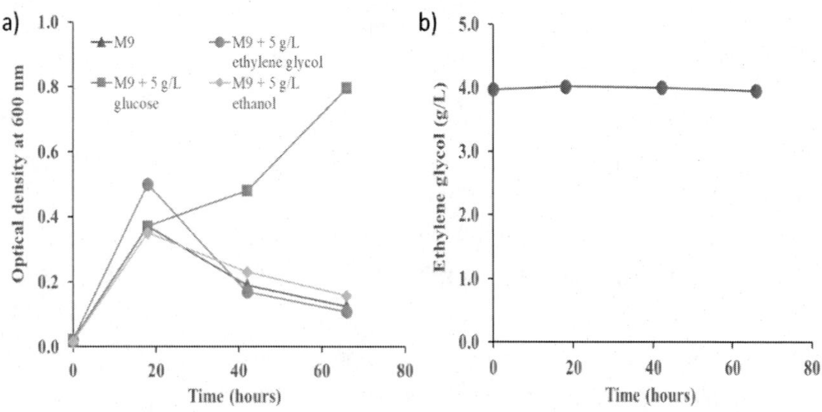

Figure B8. Assessing the ability of *B. subtilis* wildtype cells to utilize alternative substrates in M9 with 1 g/L yeast extract base medium at 37°C. a) *B. subtilis* could not grow on ethylene glycol in M9 base medium, b) Correspondingly, there was no uptake of ethylene glycol.

The substrate scope experiment described above was repeated for *B. subtilis* in M9 base medium with 1 g/L yeast extract at 37°C to confirm that temperature did not play a part in hampering the uptake of ethylene glycol. Results summarized in Figure B8a confirmed that B. subtilis could not grow on ethylene glycol given that the optical density achieved in M9 base medium with 5 g/L ethylene glycol was not significantly higher than that obtained in M9 base medium. This lack of growth on ethylene glycol could be attributed to the inability of B. subtilis to uptake ethylene glycol (Figure B8b). Deficiency in uptake of ethylene glycol may not be due to lack of transporters, but rather, the cell is intrinsically unable to utilize ethylene glycol, which, hence could not set up a driving force for the uptake of the alternative substrate.

Existence of metabolic defect in *Bacillus subtilis* BS1A1 leading to its poor growth in M9 + 1 g/L yeast extract medium thus motivates experiments seeking to examine if growth in a rich medium such as LB would activate genes that help the bacterium in using ethylene glycol and ethanol. To support this endeavor, temperature effect studies were conducted to decipher the optimal growth temperature of *B. subtilis* BS1A1. LB medium served as the medium for this set of experiments that reveal interesting results and highlight an important tradeoff in selecting an optimal growth temperature. Specifically, *B. subtilis* was able to achieve very rapid growth at 37°C compared to 30°C, but there was also attendant rapid decline in optical density after the stationary phase has been reached (Figure B9). This could arise due to the differentiation of subpopulations of *B. subtilis* into the cannibalism state due to relative lack of nutrients at stationary phase. By lysing a subpopulation of susceptible cells, nutrients could be released to support the maintenance of the residual population [82, 83]. Thus, 30°C may be a better temperature for cultivating *B. subtilis* given that a larger residual cell population could be obtained after 24 hours of cultivation, which is a time scale relevant for biotechnology considerations.

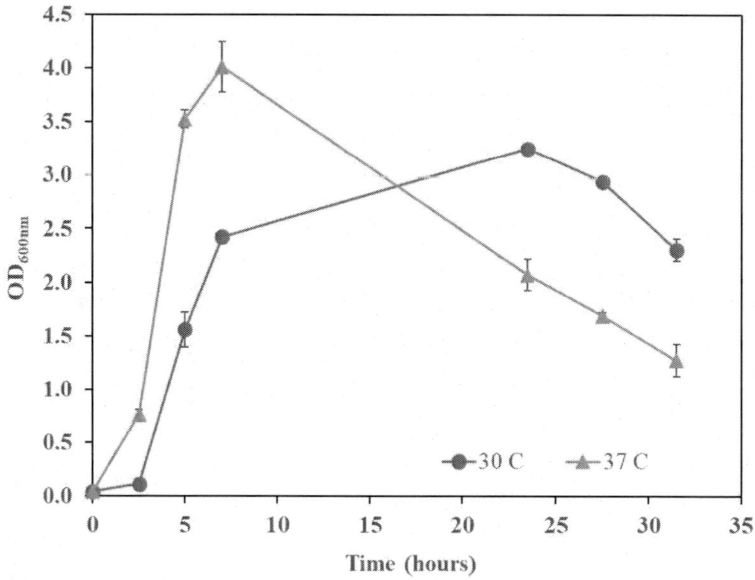

Figure B9. Comparison of growth performance of *B. subtilis* in LB medium at 30°C and 37°C reveals that *B. subtilis* grew faster at 37°C, but its cell population is vulnerable to drastic decline in optical density induced probably by differentiation into the cannibalistic state.

With the optimal growth temperature of *B. subtilis* in hand, work commenced on examining if cultivating in rich LB medium supplemented with the various alternative carbon sources would engender additional growth on top of that derived from the base medium. Experimental results revealed that, indeed, growth in a rich medium ameliorated a critical metabolic defect that prevented *B. subtilis* from better utilization of glucose for growth (Figure B10a). In particular, glucose was completely utilized within one day to support additional growth (Figure B10b). On the other hand, there remains no additional growth that could be attributed to ethylene glycol or ethanol (Figure B10a). Specifically, there was no utilization of ethylene glycol as detected by HPLC analysis (Figure B10b). Finally, there was consumption of ethanol, but no growth attributable to this substrate, which suggests that the metabolic pathway leading from ethanol was not connected to a growth promoting intermediate in central carbon metabolism.

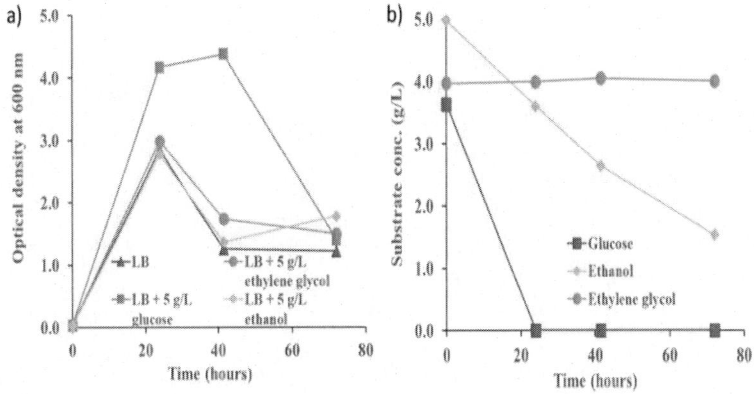

Figure B10. Assessing the substrate scope of wildtype *B. subtilis* in LB medium at 37°C. a) *B. subtilis* could not grow on ethylene glycol, but regained fitness in using glucose for growth in LB base medium, b) Glucose and ethanol were taken up by the cells, but not ethylene glycol.

5.2. Could Glycerol or Complete Supplement Mixture (CSM) Trigger Uptake of Ethylene Glycol?

Work from a colleague in the lab that shows small amount of glycerol could stimulate the uptake and catabolism of ethylene glycol in *Escherichia coli* motivated my work on examining if glycerol could indeed stimulate uptake of ethylene glycol by *Bacillus subtilis*. Experiments conducted in LB medium supplemented with 0.5 vol% of glycerol and different carbon sources with incubation at 30°C and 225 rpm revealed that glycerol could not stimulate the uptake of ethylene glycol (Figure B11b). In addition, there was no additional growth in LB supplemented medium that could be attributed to ethylene glycol (Figure B11a). Similarly, presence of glycerol could not activate genes that helped *B. subtilis* to grow on ethanol (Figure B11a). Specifically, there was uptake of ethanol by *B. subtilis*, but no growth on the substrate. This indicates that glycerol utilization could not activate genes that help with utilization of ethanol.

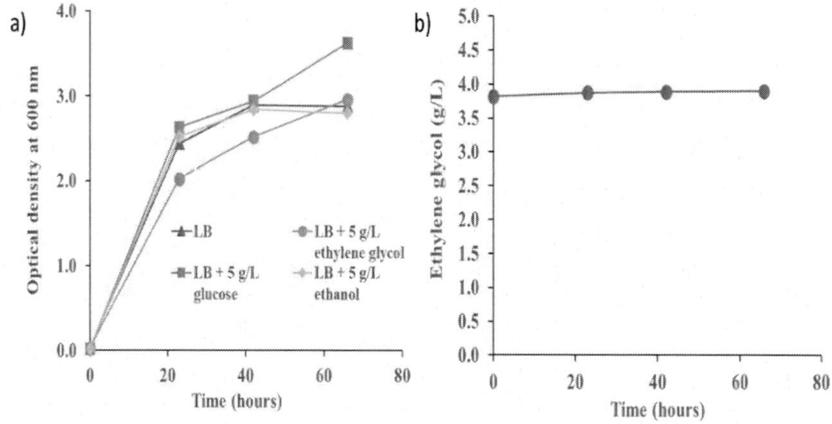

Figure B11. Assessing if glycerol could stimulate the uptake of ethylene glycol in *B. subtilis* in LB base medium at 30°C and 225 rpm. a) Growth performance of B. subtilis in LB medium supplemented with different carbon sources, b) Ethylene glycol utilization profile.

Given poor growth of *B. subtilis* BS1A1 in M9 medium supplemented with 1 g/L yeast extract, work was initiated in using complete supplement mixture (CSM) as further supplement in M9 medium with 1 g/L yeast extract in tests assessing the capability of the bacterium in utilizing a variety of different carbon substrates. Experimental results revealed that provision of additional 0.7 g/L of CSM did not improve the growth of *B. subtilis* in M9 base medium (Figure B12a). Growth remained poor even in M9 medium supplemented with 5 g/L glucose, which indicated that provision of more amino acids did not ameliorate the metabolic defect observed in *B. subtilis*. This suggests that *B. subtilis* was not deficient in amino acid acquisition, but rather another growth promoting compound was in low supply. Such a nutrient may be more readily available in higher concentrations of yeast extract as observed in growth data of *B. subtilis* in LB medium, which has 5 g/L of yeast extract.

Similar to previous experiments at 30°C and M9 base medium, *B. subtilis* could not grow on ethanol and ethylene glycol. More importantly, there was no utilization of ethylene glycol (Figure B12b). Ethanol, on the other hand, was consumed, but could not support additional growth. This highlights that provision of additional amino acids could not activate genes that could connect ethanol to central carbon metabolism. Overall, supplementation of M9 medium with 5 g/L yeast extract may be the way forward in formulating a medium capable of supporting higher biomass density of *B. subtilis*.

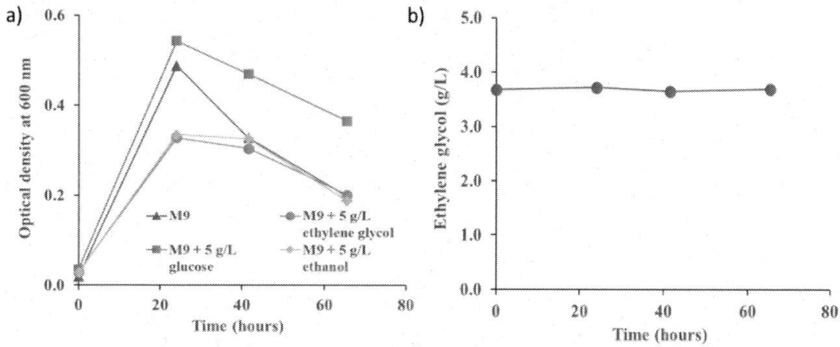

Figure B12. Assessing the substrate scope of *B. subtilis* wildtype cells in M9 with 1 g/L yeast extract and 0.7 g/L complete supplement mixture at 30°C and 225 rpm. a) Provision of complete supplement mixture did not improve growth performance across all carbon substrates, b) No uptake of ethylene glycol was observed even with provision of complete supplement mixture.

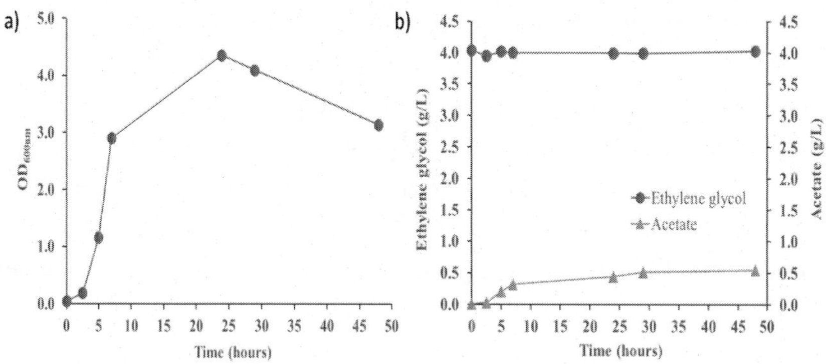

Figure B13. Confirmatory experiment in LB medium with 0.5 vol% glycerol to assess if glycerol could stimulate the uptake of ethylene glycol in *B. subtilis*. a) *B. subtilis* was able to achieve good growth in LB medium with 0.5 vol% glycerol largely due to the nutritional value of glycerol and tryptone, b) Glycerol could not stimulate the uptake of ethylene glycol.

Given the importance of the hypothesized effect of glycerol in enabling ethylene glycol uptake, experiments examining the effect were repeated in LB medium with 0.5 vol% of glycerol (Figure B13). Results again indicated that there was no uptake of ethylene glycol (Figure B13b); thereby, highlighting the importance of setting up a driving force for the catabolism of ethylene glycol in enabling the uptake of the substrate. But, good growth of *B. subtilis* in LB medium supplemented with 5 g/L ethylene glycol and 0.5 vol% glycerol indicated that ethylene glycol did not pose a toxicity issue to the cells (Figure B13a). Growth of the bacterium in the supplemented medium largely came from tryptone and glycerol.

Similar to the experiments involving glycerol and ethylene glycol, experiments examining if glycerol could activate gene expression patterns conducive for connecting ethanol to central carbon metabolism was performed in LB medium supplemented with 0.5 vol% glycerol. Experimental results revealed that there was continuous uptake of ethanol from the start of the fermentation (Figure B14a). This indicated that there was no carbon catabolite repression effect from glycerol on the ethanol catabolism pathway in *B. subtilis*. But, comparison of the data with those from LB supplemented with glycerol and ethylene glycol revealed close congruence of the growth curves, which provided conclusive evidence that there was no growth on ethanol (Figure B14b). While disappointing, the results nevertheless meant that *B. subtilis* is a good chassis for the implementation of ethanol utilization pathway connecting ethanol to acetyl-CoA. Similar to the case for *B. subtilis* cultivated in LB with glycerol and ethylene glycol, growth of the bacterium in LB supplemented with glycerol and ethanol likely came from glycerol and tryptone.

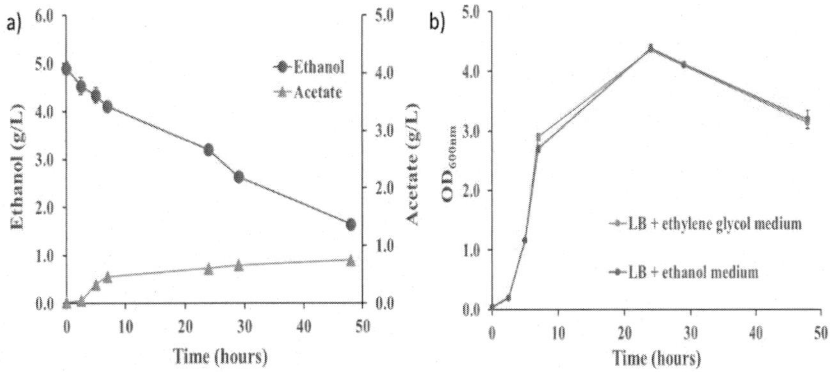

Figure B14. Assessing if glycerol provision in LB medium activated genes in *B. subtilis* that afforded growth on ethanol as alternative substrate. a) Substantial uptake of ethanol by *B. subtilis* was detected, b) But, similar growth curve for *B. subtilis* cultivated in LB ethylene glycol medium and LB glycerol medium revealed that glycerol could not stimulate *B. subtilis* to grow on ethanol.

5.3. Formulation of a Semi-Defined Medium for B. Subtilis

As mentioned early, a minimal salts based medium capable of supporting good growth of *Bacillus subtilis* is needed to help elucidate the growth potential of different alternative substrates such as ethylene glycol and ethanol. Given the time needed to prepare a trace element solution for use in formulating minimal salts medium, yeast extract was sought as a replacement of the trace element solution. The idea is theoretically feasible given that yeast extract should provide all the essential nutrients, trace elements, vitamins and growth factors necessary for growth of microorganism. In addition, yeast extract is a common component of many complex media. But observations of poor growth of a glucose supplemented M9 medium with 1 g/L yeast extract suggest that increasing the yeast extract concentration in the formulation to 5 g/L may help ameliorate the problem.

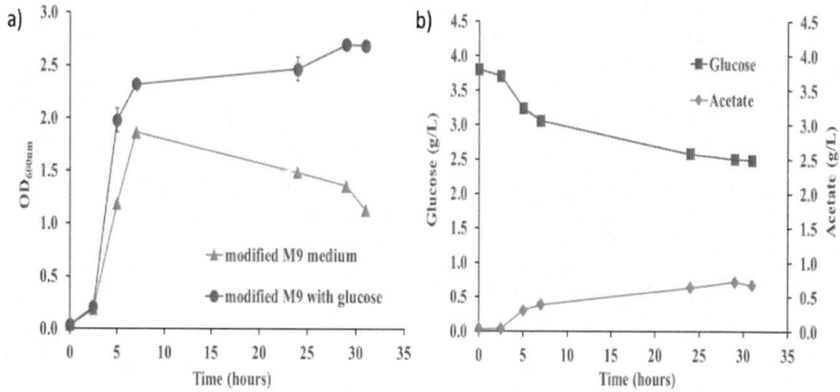

Figure B15. Assessment of growth performance from *B. subtilis* cultivated in modified M9 medium with and without 5 g/L glucose supplementation. a) Good growth was obtained for *B. subtilis* in modified M9 medium which has 5 g/L of yeast extract, b) Utilization of glucose and corresponding higher maximal growth in modified M9 medium with glucose supplementation showed that *B. subtilis* could utilize glucose in the new semi-defined medium.

Experiment tests of the growth potential of the new semi-defined formulated medium (i.e., modified M9 medium) shows that wildtype *B. subtilis* could achieve maximal optical density of 1.8 without an exogenous carbon source in the medium. This indicated that the new formulated medium has largely ameliorated the metabolic deficiency in *B. subtilis* during growth in minimal salts medium. But, the crucial test comes with the cultivation of the bacterium in the modified M9 medium supplemented with 5 g/L glucose. *B. subtilis* achieved higher maximal optical density of 2.3 in medium supplemented with glucose, which showed that the bacterium could effectively utilize supplemented glucose in the new medium (Figure B15). Experiment results confirmed that higher concentration of yeast extract in M9 medium could ameliorate the metabolic defect observed in *B. subtilis* during growth in minimal medium that presumably arise due to insufficient provision of key nutrients which needs to be identified.

But, yeast extract is a complex mixture of amino acids, peptides, vitamins and growth factors, which could confound analysis.

To remove the confounding effect of yeast extract in interpreting the ability of different carbon substrates in supporting growth, attempts were made in growing *B. subtilis* wildtype cells in M9 minimal salts medium with and without 0.7 g/L complete supplement mixture (CSM). Both media were supplemented with 5 g/L glucose. Results indicated poor growth of the bacterium in both media, which suggests that amino acids may not be the limiting component in the growth medium (Figure B16). Comparisons with other minimal salts medium formulation for *B. subtilis* revealed the noted absence of citrate and succinate in the current M9 medium formulation. Hence, it is apparent that citrate and succinate may be needed as supplementary carbon sources for a minimal salts medium. To understand this phenomenon, a RNA-seq transcriptome for *Bacillus subtilis* during exponential growth was enlisted to help pinpoint the exact reasons why the tricarboxylic acid cycle (TCA) intermediates, citrate and succinate are needed as supplements to a minimal salts medium.

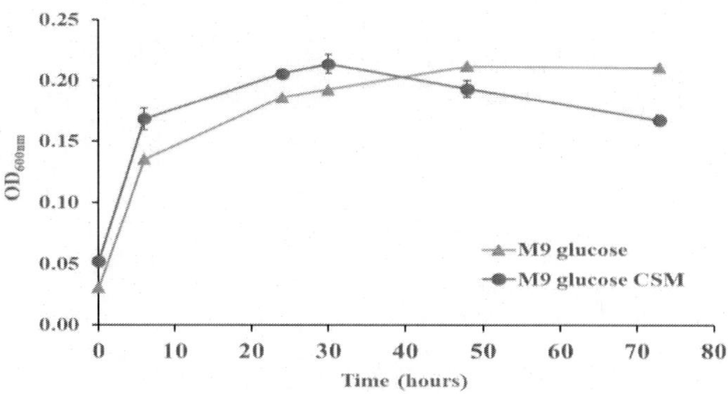

Figure B16. Comparison of growth performance of *B. subtilis* wildtype cells in M9 minimal salts medium with 5 g/L glucose supplemented with and without 0.7 g/L complete supplement mixture (CSM) revealed poor growth of the bacterium in both media; thereby, highlighting that amino acids is not the limiting nutrient for growth of *B. subtilis*.

Examination of the expression level of three citrate synthase genes, *citA*, *citZ*, and *mmgD* revealed low expression level that likely resulted in the cell being unable to efficiently synthesize citrate to help turn the TCA cycle (Table B3). More importantly, the *Bacillus subtilis* genome harbor many citrate transporter genes, which dovetail with the hypothesis that the bacterium needs to constantly uptake citrate from the environment to help fuel a metabolism deficient in citrate synthesis. Specifically, the expression level of these citrate transporters are relatively low, where most of the transporters belong to the iron-dicitrate ABC transporter family. The most highly expressed transporter is iron-dicitrate ABC transporter (ATP-binding protein), *fecF*, which has an expression level of 57. Such an expression level may be low with respect to the need to import substantial amount of citrate into the cell, but the problem would not be severe if the efficiency of the transporter is high.

Table B3. Expression level of genes related to citrate synthesis and transport in *B. subtilis*

Gene	Function	Expression count
citA	citrate synthase I	5
citZ	citrate synthase II	2
mmgD	bifunctional citrate synthase/2-methylcitrate synthase	2
cimH	citrate/malate/H+ symporter	0
fecF	iron-dicitrate ABC transporter (ATP-binding protein)	57
fecD	iron-dicitrate ABC transporter (permease)	37
fecE	iron-dicitrate ABC transporter (permease)	13
yraO	putative citrate transporter	6
fecC	iron-dicitrate ABC transporter (binding lipoprotein)	4

On the other hand, succinate biosynthesis relies on two sequential reactions mediated by succinyl-CoA synthetase and succinate thiokinase. RNA-seq transcriptome data revealed that of the two subunits of succinyl-CoA synthetase (*sucC* and *sucD*), only one subunit (*sucC*) was expressed (Table B4); thereby, pointing to an inactive path for the biosynthesis of succinate, and a need for its supplementation in the growth medium.

Table B4. Expression level of succinate synthesis genes in *B. subtilis*

Gene	Function	Expression count
sucD	succinyl-CoA synthetase (alpha subunit)	0
sucC	succinyl-CoA synthetase (beta subunit)	47

In addition to poor growth of optical density 0.2 at 48 hrs, there was also poor consumption of glucose by *B. subtilis* in M9 minimal medium and M9 medium supplemented with CSM. This suggests that glucose was not taken up by the cells. *B. subtilis* sports three distinct glucose uptake system: (i) a glcU glucose uptake system, (ii) glucose/mannose:H^+ symporter (glcP), and (iii) a phosphoenolpyruvate dependent phosphotransferase system. While supplementing the medium with mannose may help improve glucose uptake, a larger question may be the relative lack of ATP and energy to fuel glucose uptake. One circumstantial evidence supporting the hypothesis is that, in M9 minimal medium without citrate and succinate, the TCA cycle could not turn and help yield ATP. Relative lack of ATP and energy would thus be a candidate factor responsible for the cell's inability to uptake glucose. Such quagmire thus necessitates an experimental programme seeking to understand if citrate and succinate could be the sole source of carbon supporting growth. Naturally, it would also mean that for engineering growth on alternative substrate, citrate and succinate cannot be limiting carbon sources in any formulated growth medium.

While work would continue in formulating a minimal medium for *B. subtilis*, the reality suggests that a semi-defined medium with less yeast extract may be the way forward in expediting tests of whether *B. subtilis* could grow in a variety of carbon sources in efforts to nail down which metabolic pathway to pursue for the utilization of ethylene glycol. Substrates under consideration are malate, glyoxylate, acetate and citrate. Hence, a new formulated semidefined M9 medium with 2 g/L yeast extract was used as platform for testing the ability of *B. subtilis* to use the above unconventional carbon substrates as sole carbon source during aerobic cultivation at 30°C and 225 rpm. Results indicated that *B. subtilis* could grow to an appreciable optical density of 0.9 within 7 hours of cultivation in the formulated medium without supplementation of carbon source (Figure B17). This suggests that the new formulated medium is not limiting in most nutrients except for the major carbon source. Growth was supported in the new formulated medium by carbohydrates and amino acids present in yeast extract.

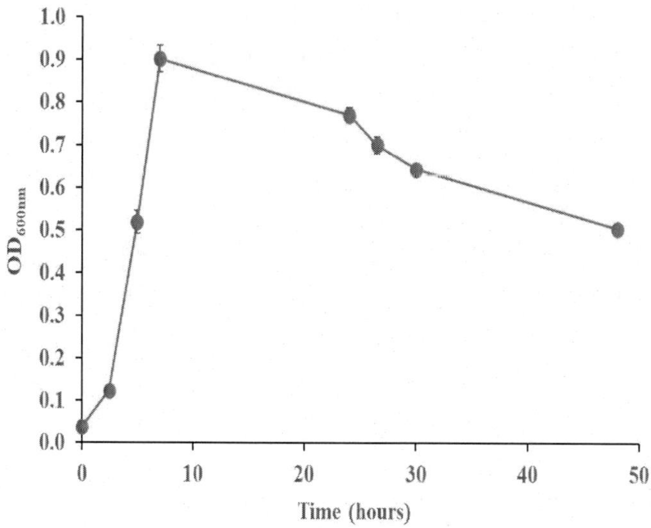

Figure B17. Supplementation of M9 medium with 2 g/L yeast extract was able to engender sufficient growth in *B. subtilis* wildtype cells; thereby, marking this semi-defined medium as possible platform for expedited testing of whether *B. subtilis* wildtype cells could grow on various intermediates in central carbon metabolism.

5.4. Assessing if Acetate Could Be Alternative Substrate for Providing Acetyl-CoA to Help Assimilate Glyoxylate Flux from Ethylene Glycol

On the other hand, experiment tests indicate strongly that *B. subtilis* could grow on acetate as a sole carbon source. This comes about from results indicating that *B. subtilis* could grow to a higher optical density of 1.3 in formulated medium with 10 g/L of acetate albeit within a longer timeframe of 26.5 hours (Figure B18a). Good consumption of acetate revealed a direct association between acetate consumption and cell growth. More importantly, the substrate utilization curve of acetate indicates that acetate consumption was activated and the gene controlling the conversion of acetate to acetyl-CoA is likely inducible by acetate (Figure B18b). Hence, all the evidence collected suggests that *B. subtilis* could grow on acetate as a sole carbon source with the substrate an evolutionarily important metabolite given the detection of an acetate inducible promoter controlling the gene responsible for its utilization. AcetylCoA synthetase is, by far, the most likely candidate for the enzyme that mediates acetate utilization. With an expression count of 12 during growth of *B. subtilis* in LB medium as measured by RNA-seq transcriptome analysis, it fits experimental data suggesting that promoter controlling the expression of acetyl-CoA synthetase is inducible.

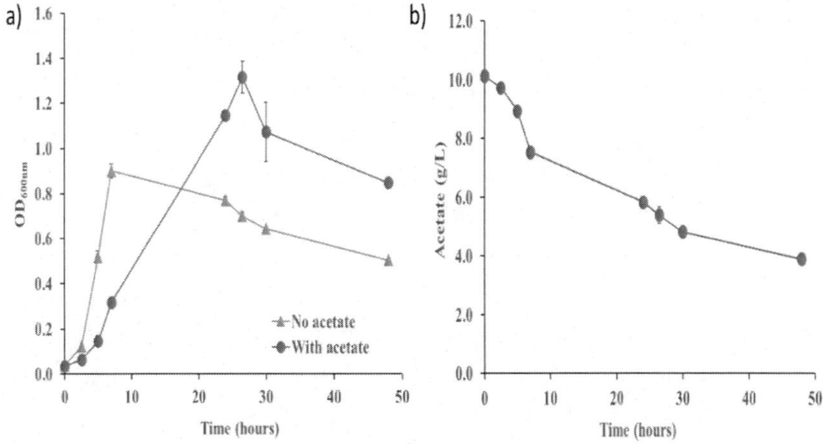

Figure B18. Assessing if *B. subtilis* wildtype cells could grow on acetate in modified M9 medium with 2 g/L yeast extract. a) Growth curve showed that *B. subtilis* wildtype was able to achieve a higher optical density with acetate supplementation compared to without supplementation, b) Acetate utilization profile revealed that the gene controlling acetate metabolism experienced inducible expression.

The result also highlights another alternative for connecting ethylene glycol to malate. Specifically, the acetyl-CoA flux for condensing glyoxylate to malate need not come from ethanol. Rather, acetate could provide the necessary flux for acetyl-CoA in a one-step reaction catalyzed by acetyl-CoA synthetase. Doing it this way reduces the number of genes that need to be expressed; thereby, helping to reduce metabolic burden of heterologous gene expression. But, the acetate to acetyl-CoA route would be tested alongside that from ethanol to acetyl-CoA to assess the relative efficacy of both pathways in providing acetyl-CoA to form malate from glyoxylate. Differences in efficacy would like arise from the differing activities of the enzymes mediating both pathways. To avoid formation of large amount of acetyl-CoA that would obscure growth from ethylene glycol, low concentration of acetate would be used in the experiments.

5.5. Selecting Appropriate Promoters for Expressing Heterologous Genes in *B. Subtilis*

Engineering *B. subtilis* with a new phenotype using heterologous expression of genes on a plasmid requires a suitable promoter. Earlier work has revealed that the autoinducible pthrC3 promoter from *Escherichia coli* could not be used in *B. subtilis* due probably to the lack of cognate inducer. But, little is known about whether the constitutive PgyrA promoter from *E. coli* could function in *B. subtilis*. Possibility exists that the promoter for the housekeeping gene, gyrA, should be conserved across bacterial species with shared regulatory elements. Hence, efforts were directed to test this hypothesis through comparing the wildtype *B. subtilis* strain with an engineered one harbouring a plasmid with ethanol utilization pathway controlled by the PgyrA promoter (strain identified as *B. subtilis* eth4). The medium chosen for the growth experiment is modified M9 medium with 5 g/L ethanol.

Results indicated that the engineered strain grew slowly in the modified M9 ethanol medium most probably due to sensitivity to ampicillin. Comparison of the maximal optical density of *B. subtilis* wildtype strain with *B. subtilis* eth4 indicated that the engineered strain could not grow on ethanol (Figure B19a). This finding is corroborated by substrate utilization profile that shows that the engineered strain did not exhibit enhanced ethanol utilization, which revealed that the ethanol utilization pathway converting ethanol to acetyl-CoA was not functional (Figure B19b). Only one reason could holistically explain all the presented observations: the PgyrA promoter controlling expression of the heterologous ethanol utilization pathway on the plasmid is not functional in *B. subtilis*. Hence, experiment results indicated that the PgyrA promoter from *E. coli* could not be used in *B. subtilis*.

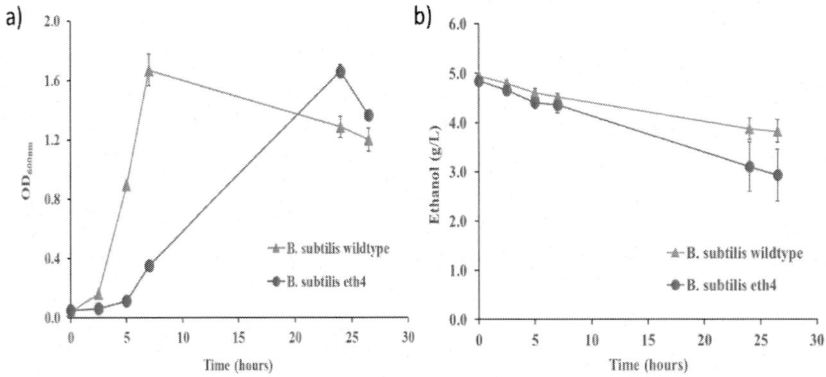

Figure B19. Determining the functional status of PgyrA from *E. coli* through assaying for growth on ethanol in strain with ethanol utilization pathway (EUP) under the control of the promoter. a) *B. subtilis* with EUP pathway under the control of PgyrA (*E. coli*) achieved the same maximal optical density as *B. subtilis* wildtype strain in modified M9 with 5 g/L yeast extract and 5 g/L ethanol at 30°C and 225 rpm, b) Ethanol utilization profile revealed that *B. subtilis* strain with EUP under the control of PgyrA could not utilize more ethanol compared to wildtype strain; thereby, indicating that the PgyrA (*E. coli*) promoter was not functional in *B. subtilis*.

The above result motivated the search for constitutive promoters in *B. subtilis* that could bind strongly to sigma factor (transcription factor) in *B. subtilis* and thus enable transcription. One candidate is the PgyrA from *B. subtilis*. Gyrase A encodes the alpha subunit of the important housekeeping enzyme gyrase A from *B. subtilis*. Thus, its expression should be constitutive, and ideally suited for our purpose of expressing heterologous genes for mediating unconventional substrate utilization. Transcriptome analysis of *B. subtilis* reveals that expression count of gyrase A gene was about 7835, and this is ranked sixth amongst all genes in the bacterium. Hence, promoter controlling expression of gyrase A should deliver good promoter strength and expression level for heterologous genes. Sequence alignment of PgyrA from *B. subtilis* and that from *Escherichia coli* reveals little long range sequence similarity, which suggests PgyrA binds to a different sigma factor from that in *E. coli*, and this meant that PgyrA from *B. subtilis* is a worthwhile candidate for initial experiments aimed at building a plasmid vector for heterologous gene expression in *B. subtilis*.

To test the functionality of the PgyrA promoter from *B. subtilis*, cell growth assay was conducted in M9 medium with 2 g/L yeast extract and 5 g/L ethanol at 30°C and 225 rpm. Experimental results revealed enhanced growth of *B. subtilis* ethBsub1 with plasmid harbouring the new promoter compared to the wildtype strain without any plasmid (Figure B20a). This suggests that the ethanol utilization pathway is functional and converts ethanol to acetyl-CoA that fuels growth. Comparison of the ethanol utilization profile in both strains also revealed higher ethanol consumption in *B. subtilis* ethBsub1 relative to the wildtype strain, thereby, indicating that the ethanol utilization pathway genes are expressed, and a driving force exists in pushing ethanol flux to acetyl-CoA (Figure B20b). All the evidence collected points to a functional PgyrA promoter that successfully activated the expression of ethanol utilization genes in *B. subtilis*, thereby, allowing this bacterial species to grow on ethanol, an ability it was not endowed with by evolution.

Overall, preliminary work has identified critical metabolic deficiency in *B. subtilis* that precluded its ability to achieve good growth in minimal medium devoid of citrate and succinate. This metabolic defect has been validated by evidence obtained from RNA-seq transcriptome of *B. subtilis* that indicated poor expression of relevant genes in rich LB medium. Hence, critical growth experiments to determine if *B. subtilis* could grow on ethylene glycol or other intermediates along the postulated heterologous pathway would need to be conducted in M9 medium supplemented with citrate and succinate or yeast extract. To expedite experimentations given the time involved in procurement of succinate, M9 medium with 2 g/L yeast extract that afford a maximal optical density of 0.9 over 7 hours may be a viable temporary option. Growth experiments also indicated that acetate could be a viable option for provision of acetyl-CoA in the ethylene glycol to malate route in place of ethanol; thereby, opening up another metabolic option for a co-feeding strategy in ethylene glycol utilization. Finally, successful demonstration of PgyrA promoter from *B. subtilis* in activating ethanol utilization pathway genes and enabling growth in *B. subtilis* meant that a promoter useful for controlling gene expression in *B.*

subtilis has been found, and this could form part of an expression vector for heterologous genes in the bacterium.

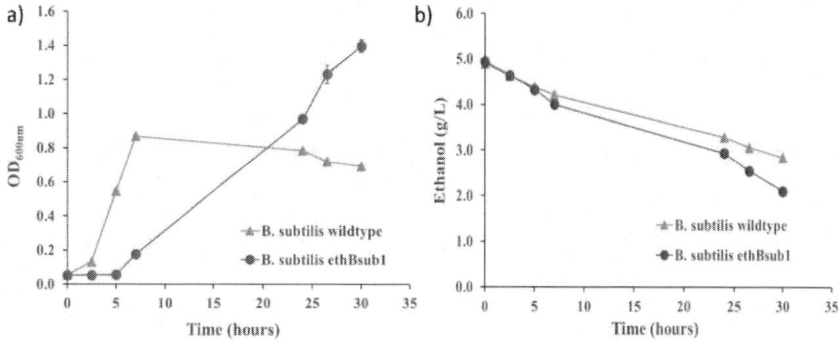

Figure B20. Functional test of the PgyrA (*B. subtilis*) cloned from *B. subtilis* through ethanol utilization and growth support test in modified M9 medium with 2 g/L yeast extract and 5 g/L ethanol at 30°C and 225 rpm. a) Higher maximal optical density in *B. subtilis* strain with EUP under the control of PgyrA (*B. subtilis*) suggests that the promoter is functional in driving EUP pathway gene expression and ethanol utilization, b) Ethanol utilization profile suggests higher ethanol uptake in strain with EUP under the control of PgyrA (*B. subtilis*) that again indicates that the promoter is functional.

6. Future Work

In substrate utilization research, a chemically-defined medium able to support good growth of the organism would greatly facilitate the work. Such a chemically defined medium for *B. subtilis* is available in the literature, and I am awaiting for the delivery of succinate to complete the medium formulation. Initial tests for the growth performance of such a chemically defined medium for *B. subtilis* will provide indicators that help answer if the bacterium is only deficient in ability to synthesize citrate and succinate, that must be provided for in the growth medium. But, a chemically-defined medium is not a strict "must" for the project, given that it is always possible to compare the growth performance of the M9 base medium with one supplemented with the growth substrate. In this case, the M9 base medium could be M9 supplemented with 2 g/L yeast extract.

Nevertheless, a chemically defined growth medium for *B. subtilis* is appended below for reference (Table B5).

Table B5. Composition of chemically defined medium for *B. subtilis* (Adapted from iGEM competition notes from iGEM 2014)

Component	Amount
L-Trytophan	0.05 g/L
Ammonium citrate	2 g/L
Potassium glutamate	1 g/L
Sodium succinate	2 g/L
KH_2PO_4	8 g/L
K_2HPO_4	5 g/L
1M $MgSO_4$	2 mL
1M $CaCl_2$	0.1 mL
NH_4Cl	1 g/L
Glucose	4 g/L

With a working growth medium as platform for conducting the research envisioned in this project, the rest of the project would flow conceptually as follows. Firstly, critical need exists in examining if *B. subtilis* could grow on malate and glyoxylate as sole carbon source. Knowledge of this would help in deciding whether the envisioned pathway connecting ethylene glycol to central carbon metabolism would work or if adjustments are needed. Next, work would commence on constructing and demonstrating the pathway from ethylene glycol to malate. Growth on ethylene glycol would be used as parameter to assess the functionality of the pathway constructed. Given that flux to acetyl-CoA could be provided by either ethanol or acetate, both substrates' utility in supplementing the flux of acetyl-CoA to afford higher efficiency of converting ethylene glycol to malate and on to other biomass precursors would be studied. Upon completing study of the first metabolic route from ethylene glycol to malate, work would commence on constructing and demonstrating the route from ethylene glycol to 2phosphoglycerate. Such a metabolic route provides a useful comparison for the earlier route that connects ethylene glycol to tricarboxylic acid (TCA) cycle. Up to this stage of the research programme, all growth experiments are designed to operate under aerobic conditions, but it would also be interesting if both pathways would work under anaerobic conditions, particularly, the route from ethylene glycol to malate that relies on TCA cycle to turn to deliver carbon flux to the rest of the metabolic network. Finally, to help confirm the metabolic route traced by the carbon flux as well as identify if they are branch points that result in carbon linkage, ^{13}C carbon labelling experiments would be conducted with ^{13}C labelled ethylene glycol, ethanol or acetate. This approach has emerged to be critical in validating metabolic flux flow in many studies on synthetic pathways or metabolic rewiring. In this project, carbon labelling experiment would play a crucial role in determining the relative contribution of ethanol and ethylene glycol to the co-feeding strategy. A schematic diagram (Figure B21) illustrating the conceptual flow of the project is as shown below.

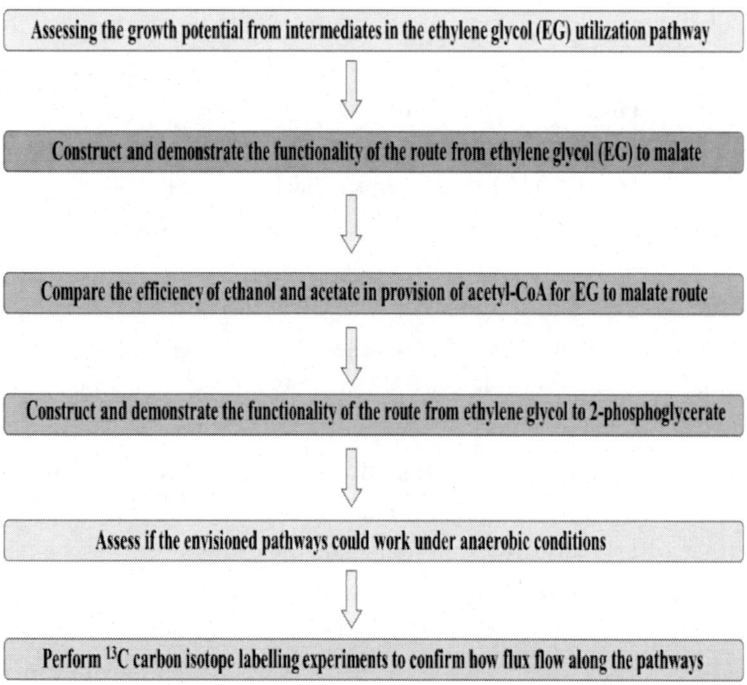

Figure B21. Schematic diagram illustrating the conceptual flow of the *B. subtilis* ethylene glycol utilization project.

Next, the pathway linking from ethylene glycol to malate would be constructed. As envisioned, one module for the pathway involves ethylene glycol utilization and connects ethylene glycol with glyoxylate. The other half of the pathway involves ethanol utilization to provide acetyl-CoA for condensing glyoxylate into malate. This module would also include the malate synthase A gene. Both pathways would be encoded as individual operon and placed under the control of the constitutive promoter, PgyrA from *B. subtilis*. Each pathway module would be encoded in a plasmid with unique replication origin and antibiotic selection marker in order to ensure compatibility in a *B. subtilis* co-transformed with both plasmids. Induction of natural competence in stationary phase *B. subtilis* would be used to transform the cells with both plasmids and the strain maintained under antibiotic selection pressure. Growth experiments with co-feeding of ethanol and ethylene glycol would be used to demonstrate the functionality of the pathway. Ethanol concentration used would be low in order to avoid supplying large amount of acetyl-CoA that would overwhelm growth delivered by ethylene glycol. Initial experiments would use an ethylene glycol and ethanol concentration of 10 g/L and 0.5 g/L respectively to demonstrate the feasibility of the approach at 30°C and 225 rpm in aerobic cultivation in a 125 mL shake flask with M9 + 2 g/L yeast extract medium or chemically defined medium (Estimated time required: 4 weeks).

After demonstration of the ethylene glycol to malate pathway, options exist to refine the pathway through assessing the relative efficacy of using ethanol or acetate to provide acetyl-CoA flux for condensing glyoxylate into malate. Ethanol could be metabolized by either the EUP pathway described in the first project of this thesis or via endogenous adhE enzyme in *E. coli*. Both options will be considered, especially the single enzyme adhE that offers pathway simplicity and flexibility. But, the alternative option of using acetate to supply acetylCoA flux may also be pursued as it brings benefit in not requiring expressing heterologous enzyme for converting acetate to acetyl-CoA. However, acetate provision in the growth medium may bring forth a reduction in pH that may not be favourable for cell growth. Plasmid encoding adhE and aceB (for malate synthase A) would be constructed and transformed into *B. subtilis* together with the ethylene glycol utilization plasmid. Growth experiments would be conducted with ethanol or acetate as co-substrate in addition to 10 g/L ethylene glycol. Similar to earlier experiments, 0.5 g/L ethanol or acetate would be provided in either M9 + 2 g/L yeast extract medium or chemically defined medium. Aerobic cultivation would be conducted at 30°C and 225 rpm in 125 mL shake flasks (Estimated time required: 4 weeks).

Completion of the above experiments marks the end of examining metabolic route 1 linking ethylene glycol to central carbon metabolism. Work would subsequently work on establishing the second route from ethylene glycol to 2-phosphoglycerate. Pathway construction necessarily require plasmids. In this case, the plasmid encoding genes for connecting ethylene glycol to glyoxylate would be reused and is designated as plasmid module 1. Module 2 would comprise genes that encode enzymes that channel flux from glyoxylate to 2-phosphoglyerate, and this would constitute plasmid 2 for the project. Both plasmids would be co-transformed into *B. subtilis* using the natural competence method. After the genetically engineered strain is obtained, growth experiments utilizing 10 g/L ethylene glycol would be conducted in either 25 mL of M9 + 2 g/L yeast extract medium or chemically defined medium in 125 mL shake flask at 30°C and 225 rpm (Estimated time required: 4 weeks).

All of the growth experiments conducted till now are aerobic experiments. But, it would be interesting to test the two constructed pathways under anaerobic conditions. Scientific interest in this aspect of the work arises from possibility that the tricarboxylic acid (TCA) cycle in *B. subtilis* may not function optimally under anaerobic conditions. This would have significant negative impacts on the ethylene glycol to malate pathway that relies on proper and efficient functioning of the TCA cycle to bring metabolic flux to the rest of central carbon metabolism. Another scenario would be anaerobic conditions may not help provide a sink for surplus NADH generated by the heterologous pathways. This is unlike the case in aerobic cultivation where respiration consume excess NADH [84]. Hence, anaerobic conditions may affect pathway functionality. Experiments with wildtype *B. subtilis* under anaerobic conditions would be conducted first as control. If successful, strains with both pathways would be tested under anaerobic growth conditions to assess the relative efficacy of both pathways in converting carbon flux in ethylene glycol into biomass. Different from aerobic growth experiments, anaerobic growth experiment would be conducted with 0.5 mL of cell-medium mixture in a 2.0 mL screw-capped HPLC glass vial. Incubation conditions would be 30°C and 225 rpm for a few days, with each vial as a sampling point. Medium used would be M9 + 2 g/L yeast extract medium or chemically defined medium (Estimated time required: 2 weeks).

Finally, ^{13}C carbon isotope labelling experiments provide a tool for tracking the flow of carbon and metabolic flux from labelled substrate into central carbon metabolism. The approach is particularly useful for determining if the ethylene glycol to malate pathway is functioning the way it is construed conceptually. Specifically, ^{13}C labelled ethylene glycol and ethanol or acetate could be used to diagnose if intracellular malate is doubly labelled for ^{13}C. If that is the case, it would mean that carbon flux did flow according to the planned route from ethylene glycol and ethanol or acetate. Other metabolites such as glycolate and glyoxylate that could be detected by ion-exchange liquid chromatography mass spectrometry could also be detected to confirm that path in which labelled carbon flow along the pathway. On the other hand, ^{13}C labelled ethylene glycol can also be used to track the carbon flow along the ethylene glycol to 2-phosphoglycerate route. Signs that the pathway is working according to plan include detection of ^{13}C labelling in glycolate, glyoxylate and pyruvate. After connecting to 2phosphoglycerate in glycolysis, the metabolic flux from ethylene glycol should also flow down to pyruvate in glycolysis. Thus, one good measure of whether the carbon flux from ethylene glycol has successfully connected and flowed to glycolysis is through detecting ^{13}C labelling in pyruvate and other central carbon metabolism metabolites. All metabolites mentioned above are detected through analyzing whole cell lysate of cells feed with the corresponding ^{13}C labelled substrates. (Estimated time required: 4 weeks inclusive of setting up the LC-MS analysis pipeline).

Overall, the estimated amount of time needed for the project is about 20 weeks. Time needed for plasmid construction has been factored in the estimates presented above. Tuning of gene expression level may need another 3 weeks. Coupled with 4 weeks allocated for general troubleshooting, the total time needed for the project is about 27 weeks. Details for the timeline of the project can be found in Table B6 below.

Table B6. Timeline for the implementation of the *B. subtilis* ethylene glycol utilization project

No.	Major tasks in the project	Time required
1	Assessing the growth potential from intermediates in the ethylene glycol (EG) utilization pathway	2 weeks
2	Construct and demonstrate the functionality of the route from ethylene glycol (EG) to malate	4 weeks
3	Compare the efficiency of ethanol and acetate in provision of acetylCoA for EG to malate route	4 weeks
4	Construct and demonstrate the functionality of the route from ethylene glycol to 2-phosphoglycerate	4 weeks
5	Assess if the envisioned pathways could work under anaerobic conditions	2 weeks
6	Perform ^{13}C carbon isotope labelling experiments to confirm how flux flow along the pathways	4 weeks
7	Possible need to tune promoter strength of ethylene glycol utilization pathway	3 weeks
8	General troubleshooting	4 weeks
	Total:	27 weeks

CONCLUSION

Need for shifting the carbon source for commodity chemicals and consumer products from fossil fuel to renewable resources has motivated the search for alternative substrates in biotechnology and metabolic engineering. Glucose has been the dominant carbon source for the generation of a variety of bioproducts over the past few decades. This comes about due to the status of glucose as the preferred carbon source for many microbial species across the tree of life, which meant that metabolic pathways are available in most species for its metabolization. Such a case is not common for many of the alternative substrates that have received some attention for pathway engineering in the past decades. Specifically, a new pathway needs to be constructed using existing enzymes from different species or entirely new enzymes to mediate the metabolization of a new alternative substrate and connect the carbon flux to central carbon metabolism.

One possible alternative substrate is ethylene glycol that could be derived from hydrolysis of lignocellulosic waste or plastic PET waste. Adequate supply of reasonably priced ethylene glycol should be available under the above scenario. But, what is more fundamental about this project is the demonstration of the feasibility of building a metabolic route from ethylene glycol to central carbon metabolism in *Bacillus subtilis* that is devoid of the glyoxylate shunt and glycerate pathways. Such a microbial chassis places an inevitable challenge on the project, but the disadvantages inherent in this challenge is modulated by the Generally regarded as safe status of *B. subtilis*.

Besides building a feasible pathway that could bring carbon flux from ethylene glycol to central carbon metabolism for building biomass, another major hypothesis tested in this project is whether different entry points in central carbon metabolism presents different efficiency in generation of biomass precursors for cell growth. Hence, two routes are planned for bringing carbon flux from ethylene glycol to central carbon metabolism. The first path connects ethylene glycol to malate in tricarboxylic acid cycle through glyoxylate, which serves as an important metabolic branch point. To help condense glyoxylate to acetyl-CoA for forming malate, an exogenous source of carbon must be supplied to provision acetyl-CoA for the purpose. Ethanol coupled with the ethanol utilization pathway explored in first project of this thesis is a viable pathway for providing acetyl-CoA. Alternatively, acetate could also be converted into acetyl-CoA through endogenous acetyl-CoA synthetase in *B. subtilis*. As construed, co-feeding of ethylene glycol and ethanol or acetate is required to connect ethylene glycol to malate. Since ethanol or acetate may have different efficacy in generating acetyl-CoA for condensing glyoxylate, both substrates would be compared to determine the preferable route for generating biomass from ethylene glycol.

The second pathway involves the sole feeding of ethylene glycol and concerns the connection of the substrate to 2-phosphoglycerate in glycolysis. Similar to the ethylene glycol to malate route, glyoxylate is the metabolic branch point that divides the first half of the pathway from ethylene glycol to glyoxylate from the second half that runs to 2phosphoglycerate. The two halves of the pathway represent distinct pathway module that could be individually encoded in separate plasmids; thereby, allowing the expression level of the pathway to be modulated to help maintain a constant flux along the entire pathway. Such modulation is necessary given that activities of the enzymes in separate pathway module may be different.

Both pathways for connecting ethylene glycol to central carbon metabolism would be tested for biomass formation under aerobic conditions. But, given that some biotechnology applications require anaerobic growth, there is practical interest in understanding the efficiency of both pathways to generate biomass under anaerobic conditions. Such applied experimentation may also yield insights in basic biology on utilization of alternative substrate under different oxygen environment. From the theoretical perspective, anaerobic conditions are expected to have an impact on the ethylene glycol to malate route given that tricarboxylic acid cycle may not function optimally under these conditions. Ethylene glycol to 2phosphoglycerate route, on the other hand, should be invariant under different oxygen tension. Finally, ^{13}C carbon labelling experiments would be conducted to track and confirm the directionality of the metabolic flux flow from ethylene glycol to central carbon metabolism. Use of whole-cell lysate and ion-exchange liquid chromatography-mass spectrometry should provide a glimpse of labelling of pathway intermediates such as glycolate, glyoxylate, malate or pyruvate for confirming that the constructed pathway has indeed channelled flux to the designated connecting molecule in central carbon metabolism.

Overall, in constructing pathway connecting ethylene glycol to central carbon metabolism using genes cloned from *E. coli*, the study hopes to demonstrate the feasibility of wiring an ethylene glycol naïve *B. subtilis* for use in biotechnology revolving tapping the carbon flux from the alternative C2 substrate. Challenges involved in this endeavour include the expression of a number of heterologous pathway genes in a host that may succumb to excessive metabolic burden, which meant that modulation of expression level is critical for experiment success. Comparison between connecting ethylene glycol to glycolysis or tricarboxylic acid cycle would provide important data indicating the relative efficiency in which glycolysis or tricarboxylic acid cycle in channelling flux for building biomass. This constitute important data that could guide our understanding in selecting appropriate entry points for connecting flux from alternative substrates. Such a choice is likely species-specific given differing relative abundances of different enzymes in central carbon metabolism in different species. But, its core essence is similar to the evolutionary forced selection of particular entry points in central carbon metabolism for the commonly used substrates such as glycerol, fructose and glucose.

Chapter 5

FUTURE PROJECT C: EVALUATING IF CONNECTING ETHYLENE GLYCOL TO ACETYL-COA IN *BACILLUS SUBTILIS* IS THE OPTIMAL APPROACH FOR POWERING UP AMINO ACID AND PROTEIN SYNTHESIS

1. INTRODUCTION

Protein synthesis is one of the major goals in earlier part of biotechnology research such as during the attempt to synthesis recombinant insulin in bacterial cell [85]. However, the subfield of improving protein synthesis has received less attention in recent years as focus shifted to ramping up the production of small molecule and complex natural products in different microorganisms through metabolic engineering. At its heart, protein synthesis is intimately tied in to cell growth processes as increase in cell volume with cellular building blocks for growth require the synthesis of biomass precursors. But, if the goal is to synthesize a recombinant protein to high titer in a cell, methods must be developed for diverting flux to synthesis of amino acids and proteins without significantly affecting cell growth rate. Such methods could be derived from rational approach utilizing our knowledge of metabolism or from trial and error experimentation. In this project, the rational approach would be adopted.

Protein synthesis is reliant upon the efficient synthesis of most of the twenty naturally occurring amino acids, which meant that flux must be diverted to several key metabolic precursors that supply carbon for synthesizing amino acids. These metabolites are oxaloacetate, α-ketoglutarate, pyruvate, 3-phosphoglycerate, and erythrose-4-phosphate. Hence, flux is diverted from glycolysis, tricarboxylic acid cycle and pentose phosphate pathway to support the synthesis of various types of amino acids needed to manufacture a protein. One may ask: given the divergent metabolic pathways involved, how to we distribute flux amongst the different arms of metabolism to help provide equal abundance of amino acids for synthesizing a target protein? The answer from the rational perspective may be acetyl-CoA.

The metabolic importance of acetyl-CoA is without doubt [86-88]. Specifically, it is the key central carbon metabolism intermediate that connects glycolysis and tricarboxylic acid (TCA) cycle. Given that glycolysis is also connected to pentose phosphate pathway, funnelling of carbon flux from a substrate to acetyl-CoA would connect at least two large metabolic branches of central carbon metabolism in glycolysis and tricarboxylic acid cycle. What is more weakly connected to acetyl-CoA is perhaps pentose phosphate pathway given the number of enzymatic steps that separate it from acetyl-CoA. By "weak connection," the concept pertains to the low availability of carbon flux to a specific metabolite in pentose phosphate pathway that has been funnelled from acetyl-CoA. Such low availability of carbon flux could be due to the diversion of carbon to form other intermediates that branched from the main metabolic pathway, thereby, leaving only relatively little flux for the target metabolite.

Protein synthesis in biotechnology workhorse is commonly built upon flux derived from glucose [89]. But, the metabolic route of glucose to pyruvate and especially, oxaloacetate and α-ketoglutarate is long and final flux to the metabolites feeding to synthesis of amino acids may be low. Hence, there is practical interest in identifying a substrate that could efficiently fuel amino acid synthesis for building recombinant proteins in biotechnology. While many common substrates such as xylose, fructose and glycerol could fuel amino acid synthesis and cell growth, but they all suffer from entry into glycolysis and an ensuing "long pathway distance" to key intermediates that feed amino acid synthesis in tricarboxylic acid (TCA) cycle. This brings forth a need to develop a new substrate that could efficiently and expeditious bring flux from substrate to acetyl-CoA to fuel amino acid synthesis reactions in glycolysis and TCA cycle. Acetyl-CoA is the preferred molecule because of its proximity to pyruvate, oxaloacetate and α-ketoglutarate.

One possible substrate that could be developed for feeding amino acid synthesis is ethylene glycol. Specifically, the carbon flux could be derived from a molecule that is the hydrolysis product of lignocellulosic waste as well as polyethylene terephthalate (PET) plastic waste. Hence, availability of substrate should not be a problem. But, what is more important is that ethylene glycol could be connected fairly expeditiously to acetyl-CoA in three steps; thereby, ensuring a high metabolic flux to central carbon metabolism to fuel the synthesis of amino acids through different intermediates. The pathway starts with the ethylene glycol conversion to glycoaldehyde through fucO (L-1,2-propanediol oxidoreductase) from *Escherichia coli*. Glycoaldehyde would subsequently be converted into acetyl-phosphate through phosphoketolase from actinobacteria. Finally, acetyl-phosphate would be converted to acetyl-CoA through phosphate acetyltransferase (pta) from *E. coli* (Figure C1). Inspiration for the glycoaldehyde to acetyl-CoA route came from the work of Lu and coworkers [90].

Microbial chassis chosen for this project is *B. subtilis* given its status as a Gram-positive biotechnology workhorse with good secretion capability [91, 92]. This latter capability can be efficiently tapped to secrete recombinant proteins synthesized in the microbe. Green fluorescent protein is selected as the model protein for this project given ease of detecting its production and relative abundance through fluorescence method that can be implemented using a microplate reader.

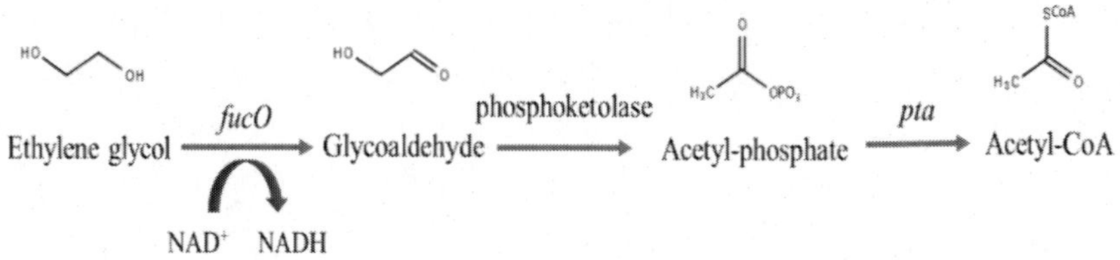

Figure C1. Architecture of the utilization pathway routing carbon flux from ethylene glycol to acetyl-CoA. Specifically, ethylene glycol would be converted to glycoaldehyde. A phosphoketolase from actinobacteria would next convert glycolaldehyde to acetyl-phosphate. Finally, acetyl-phosphate would be converted by a phosphate acetyltransferase (pta) from *E. coli* into acetyl-CoA.

Overall, a larger goal of this project is to compare the relative efficiency in which different pathways leading from ethylene glycol to central carbon metabolism help improve protein production. Specifically, the conceptual underpinnings revolves the hypothesis that if the entry point to central carbon metabolism for ethylene glycol is far away from metabolites that feed amino acid synthesis, flux diversion through side reactions may be a major factor limiting amino acid production, which is likely to have knock-on effects on protein production. To this end, pathways linking ethylene glycol to central carbon metabolism in the previous project could be modified and used for comparing the relative efficiency of different entry points to central carbon metabolism in channelling flux to amino acid synthesis. Central to the hypothesis is the postulation that different entry points have different distance to metabolites feeding amino acid synthesis resulting in different carbon flux for protein production. Altogether, three pathways with different entry points into central carbon metabolism would be tested: (i) ethylene glycol to malate, (ii) ethylene glycol to acetyl-CoA, and (iii) ethylene glycol to 2-phosphoglycerate. Knowledge gained from these experimentations in combination with theoretical understanding would provide useful guidance for the most suitable metabolic entry point for ethylene glycol with respect to improving protein production in *B. subtilis*.

2. Theoretical Underpinnings of the Project

In this project, three different metabolic pathways linking ethylene glycol to intermediates in central carbon metabolism would be compared for their efficiency in channelling flux for amino acid and protein synthesis. The three pathways end at different intermediates in central carbon metabolism: malate, acetyl-CoA, and 2-phosphoglycerate. Such a setup would likely have implications for the amount of metabolic flux available to intermediates that feed into amino acid synthesis. For example, flux entering central carbon metabolism at malate may be better diverted to amino acid synthesis at oxaloacetate and αketoglutarate compared to flux entering at 2-phosphoglycerate that has to travel some distance across a few intermediates to feed oxaloacetate and α-ketoglutarate. This imbalance in metabolic flux available at each metabolite that feeds amino acid synthesis may have an effect on protein production in strains with different pathways. Details of the different pathways studied in this project would be described next.

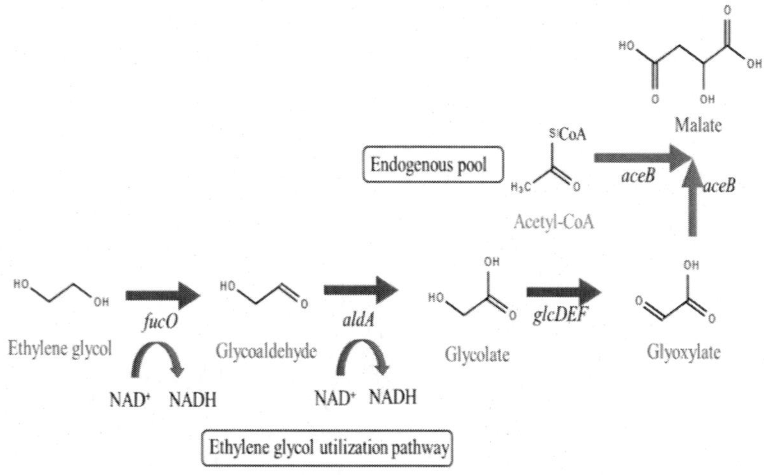

Figure C2. Pathway guiding carbon flux from ethylene glycol to malate. Different from the co-feeding strategy utilized in a previous project, the current work relies on the endogenous pool of acetyl-CoA to condense glyoxylate from ethylene glycol utilization pathway for forming malate.

Ethylene glycol to malate is the first pathway under examination in this project. It pushes carbon flux from ethylene glycol along the path marked by glycoaldehyde, glycolate and glyoxylate to malate. All enzymes are cloned from *E. coli* for the purpose. The first step is the NAD^+ dependent conversion of ethylene glycol to glycolaldehyde powered by fucO. Subsequently, glycoaldehyde is converted to glycolate by an aldehyde dehydrogenase, aldA. This step generates a NADH. Next, glycolate is converted to glyoxylate by glycolate dehydrogenase (glcDEF). Finally, a molecule of glyoxylate would condense with one molecule of acetyl-CoA to form malate that enters central carbon metabolism. Overall, the pathway generates two NADH per turn of the cycle. But, its efficiency may be limited by the size of the endogenous pool of acetyl-CoA available. As a TCA cycle intermediate, malate is close to oxaloacetate and α-ketoglutarate, but flux from the intermediate may have difficulty travelling to pyruvate and 3-phosphoglycerate to feed the synthesis of other amino acids. The pathway would be encoded on one plasmid with expression controlled by PgyrA from *B. subtilis*.

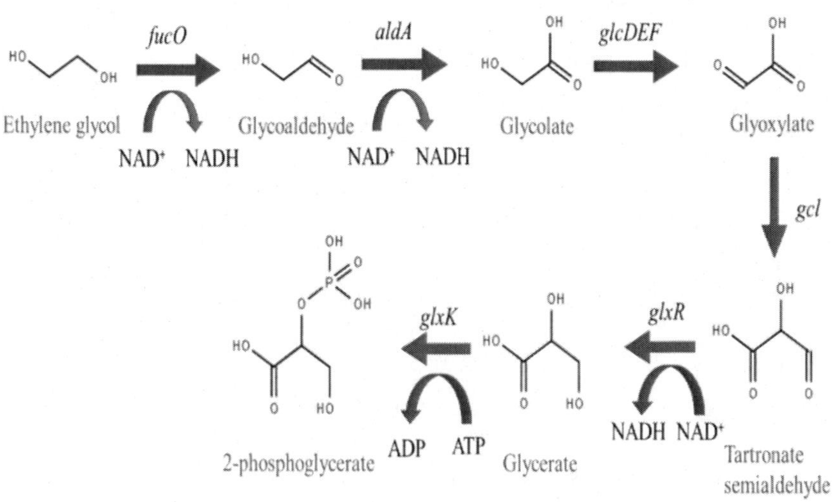

Figure C3. Architecture of pathway leading from ethylene glycol to 2-phosphoglycerate. The first part of the pathway is similar in structure to the one from ethylene glycol to malate, but it has a branch from glyoxylate to 2-phosphoglycerate.

The second pathway guides ethylene glycol to the glycolysis intermediate, 2phosphoglycerate, in 6 enzymatic steps. First half of the pathway from ethylene glycol to glyoxylate would be encoded on a plasmid. This is followed by the second half of the pathway from glyoxylate to 2-phosphoglycerate. Specifically, glyoxylate is converted to tartronate semialdehyde via glyoxylate carboligase. Next, tartronate semialdehyde would be converted by tartronate semialdehyde reductase to glycerate. Finally, glycerate-2-kinase converts glycerate to 2-phosphoglycerate. All genes for enzymes in the pathway are from *E. coli*. Overall, the pathway consumes 1 ATP and generates 3 NADH per turn of the cycle. With entry to 2-phosphoglycerate along the glycolysis pathway, the flux has easy access to pyruvate, 3phosphoglycerate and pentose phosphate pathway for synthesis of amino acids. However, 2phosphoglycerate may not be able to deliver much flux to oxaloacetate and α-ketoglutarate. This may create an imbalance in synthesis of different types of amino acids that may later manifest as constraints to protein production.

The final pathway links ethylene glycol to the growth promoting metabolite, acetylCoA in three steps as described in the Introduction (Figure C1). Overall, the pathway generates one NADH in the first step, and reduces a CoA and inorganic phosphate for function. As such, efficiency of the pathway may be dependent on the supply of Coenzyme A. Amongst all three investigated pathways, this pathway offers the shortest route to central carbon metabolism, and enters at a strategic molecule, acetyl-CoA, located at the connection between glycolysis and tricarboxylic acid (TCA) cycle. This location in the metabolic network affords acetyl-CoA an opportunity to more evenly distribute the flux to intermediates that feed amino acid synthesis in glycolysis, pentose phosphate pathway and TCA cycle. Hence, this pathway may provide the highest efficiency in protein production and green fluorescent protein fluorescence intensity. This pathway would be encoded on one plasmid in which PgyrA controls the expression of the pathway genes. Gene encoding green fluorescent protein would be encoded on another plasmid controlled by the same promoter, PgyrA.

Cell growth would be used as driving force to pull the flux from ethylene glycol into central carbon metabolism given that all three pathways terminate at intermediates in the metabolic network. Thus, cellular building blocks would be used for building both green fluorescent protein and biomass. Amount of green fluorescent protein synthesized could be correlated with intensity of fluorescence at 545 nm in a microplate reader with excitation at 395 nm. Hence, fluorescence intensity per unit OD would be a good metric for assessing the relative efficiency in which different pathways would provide flux to support amino acid synthesis and protein production.

3. Literature Review

Production of recombinant protein via genetic engineering approaches in bacteria was first started in the 1970s. Ever since, much effort has been expended in optimizing various aspects of gene expression and cell cultivation for improving protein production or expressing hard to produce protein. More recently, strain engineering has been investigated for improving recombinant protein production in bacteria [93]. Expression tuning is perhaps the foremost approach for improving recombinant protein production, and involves the optimisation of promoter, ribosome binding site, expression vector, inducer and codon optimization of encoding gene [94]. Moving beyond vector design and optimization of cell culture medium and conditions, systems level analysis has been employed in devising strategies for improving protein production [95]. Such approaches have utilized metabolic engineering perspective to identify rate-limiting steps of protein production, as well as examining ways to divert flux towards amino acid synthesis [95]. More importantly, attempts were also made to divert cellular resources for biomass formation towards protein production given that protein production is an essential process of biomass formation [95]. To do this, multiple genetic changes such as promoter tuning, gene expression and gene deletion must be implemented at the genome level, and such efforts ushered in the age of genome engineering for improving recombinant protein production [95]. Improvements in fermentation concepts have also demonstrated the feasibility of using two stirred tanks to separate biomass formation and protein production [96]. Generally, the goal in recombinant protein production has been the production of large amount of cells, prior to inducing the expression of heterologous proteins [97]. Besides upstream strategies described above, there has also been research activities aiming to improve recombinant protein production through utilizing next-generation bioprocessing approaches and methodologies [98]. These will not be of focus in this literature review.

The topic of improving recombinant protein production has more recently touched on the topic of food sustainability, and more broadly, food security. Specifically, recombinant protein production holds possibility for producing food protein in microbial cells [99], which provides an alternative protein source for the world's population and improving our food security. For example, a recent study modelled the upstream economic feasibility of recombinant production of food protein, Bovine Alpha Lactalbumin. Examination of substrate utilization profiles and productivities highlight its economic feasibility [99].

Bacillus subtilis is a well-studied industrial biotechnology workhorse useful for a variety of applications such as over-production of recombinant protein or secretion of metabolites and protein [100-102]. Besides intracellular expression of recombinant protein, there has also been attempts at extracellular expression of recombinant proteins in *B. subtilis* using dual promoter system [103]. In this project, green fluorescent protein (GFP) is used as a reporter for understanding the relative production levels of protein through fluorescence intensity measurement. Green fluorescent protein has been successfully expressed in *B. subtilis*, and has emerged as an important tool for probing bacterial cell biology [104]. More important for biotechnology applications, the fluorescent from GFP is stable. As a well-studied fluorescent protein with different variants suited for different applications, studies have been conducted to illuminate the best performing green fluorescent protein variant for *B. subtilis*. Experimental results reveal that super-folder GFP is dimmer than GFP in *B. subtilis*

[105]. Such a result supports the choice of GFP as marker protein in this project. But, GFP expression is also known to be heterogenous at the single cell level, with population level variance in expression level and brightness of fluorescent protein [106]. One experimentally validated approach to remedy the problem is to express GFP under the control of CodY promoter with R214C mutation, which provides more homogenous expression of the fluorescent protein [106].

Heterologous protein production in bacteria taps on the flux available in central carbon metabolism, particularly those that feeds off pyruvate, 3-phosphoglycerate, oxaloacetate, αketoglutarate, erythrose-4-phosphate and pentose phosphate pathway. Hence, to increase protein production, one must improve carbon flux to amino acid precursors most commonly through metabolic engineering. The target is naturally the complex intertwined network of cellular reactions known as central carbon metabolism [107]. Comprising glycolysis, gluconeogenesis, tricarboxylic acid cycle (TCA), and pentose phosphate pathway, engineering central carbon metabolism requires careful modulation of flux as the cell needs a reasonable pool (concentration) of all major intermediates in the metabolic network to survive. To avoid locking *B. subtilis* into distributing carbon and energy between cell maintenance and biomass formation, culture conditions should not be carbon limited [108]. This informs experiment design to provide ethylene glycol in some excess to help supply flux for green fluorescent protein synthesis.

One clear possibility for improving flux for protein production would be to increase the feeding off of carbon flux to amino acid synthesis reaction. But, this may create metabolic imbalance where pool of important metabolites in the cell are depleted. Given that tuning of flux at a granular level is experimentally laborious, metabolic flux analysis using *in silico* simulation techniques may provide clear indications for the reaction requiring tuning [109]. Such an approach enjoy advantages of a global view with predictions that are accessible for experimental verification given their smaller solution space compared to the initial problem.

But, the larger question is whether tuning flux delivery to particular metabolic nodes that supply amino acid synthesis is sufficient for improving heterologous protein synthesis. One factor not considered till now is regulatory protein factors that controls the partition of energy and carbon between cell maintenance and protein production. Such protein has been known to mediate the division of flux between central carbon metabolism and peptidoglycan synthesis [110], but relatively little is known about regulatory factors that control the overall flux availability for protein synthesis. From theoretical reasoning, protein production may not be amenable to centralized control by a global regulator given the distributed nature in which different amino acids are synthesized by tapping flux from metabolites in different branches of central carbon metabolism. In addition, tuning expression levels of global regulatory factors may be a blunt approach for engineering over-expression of proteins, and may have untoward physiological effects that counteract protein over-expression. However, it is not correct to think that such tuning of global regulatory factors such as transcription factors could not increase protein yield. For example, a recent study that modulated global nitrogen and carbon metabolism through titrating the levels of transcription factor CodY and CcpA improves protein yield by 290%; thereby, demonstrating the utility of the approach [101].

Such global regulatory factors also have effects on the choice of supplementary carbon sources (if need arise) for fermentation. Specifically, in *B. subtilis*, CodY and CcpA act in concert to repress the use of alternative carbon sources if glucose is present [111]. This meant that glucose cannot be used as supplementary carbon source if the objective is the utilization of ethylene glycol. This then point the thinking towards modulating the levels of global regulatory factors as a means to improve protein production. A comprehensive study detailing the regulatory roles of the global regulatory factor, CodY, reveals increases in branched chain amino acids and intermediates in arginine, proline and glutamate metabolism together with reduction in pyruvate and glycerate as CodY activity decreases [112]. Thus, modulating the activity level of

CodY may be one approach for improving protein production in later part of the research if time permits.

Overall, the above literature review has highlighted, to the best of my understanding, no work has been done in channelling ethylene glycol flux to synthesis of a recombinant protein such as green fluorescent protein. Furthermore, the pathways directing carbon flux from ethylene glycol to different metabolites in central carbon metabolism have not been implemented in *B. subtilis*, even though they have been inspired in part by endogenous pathways in *E. coli*, and in the case for ethylene glycol to acetyl-CoA, by the literature. The main hypothesis examined in this proposed work that posits that different entry point for carbon flux in central carbon metabolism may lead to different availability of amino acids for synthesizing green fluorescent protein is a relatively new concept that again to the best of my literature search ability, has not been examined before. Such a hypothesis would naturally place less emphasis on choosing a novel recombinant protein for expression, and a proven fluorescent protein was selected for heterologous expression to serve as readout of the relative flux diverted from ethylene glycol to support protein synthesis.

4. Future Work

The conceptual flow of the ethylene glycol to protein synthesis project is as outlined in Figure C4. First part of the project concerns the establishment of the green fluorescent protein expression system in B. subtilis. To this end, the green fluorescent protein from Aequorea Victoria (jellyfish, UniProtKB-P42212) would be expressed via a plasmid vector with expression under the control of PgyrA from *B. subtilis*. Time course experiment to evaluate the timing of green fluorescent protein expression in *B. subtilis* would be conducted in growth assays in LB and M9 + 2 g/L yeast extract medium at 30°C and 225 rpm in 125 mL shake flask. Different growth medium was used to establish the effect of different nutritional environment on the maximum fluorescence intensity level (and protein production level) that could be achieved in *B. subtilis*. On the other hand, time profile experiment would provide insights into when expression of green fluorescent protein would be initiated which would inform the timing of maximal fluorescence intensity and protein production level. Green fluorescent protein would be encoded on a plasmid with PgyrA as promoter with an appropriate antibiotic resistance marker (Estimated time required: 4 weeks).

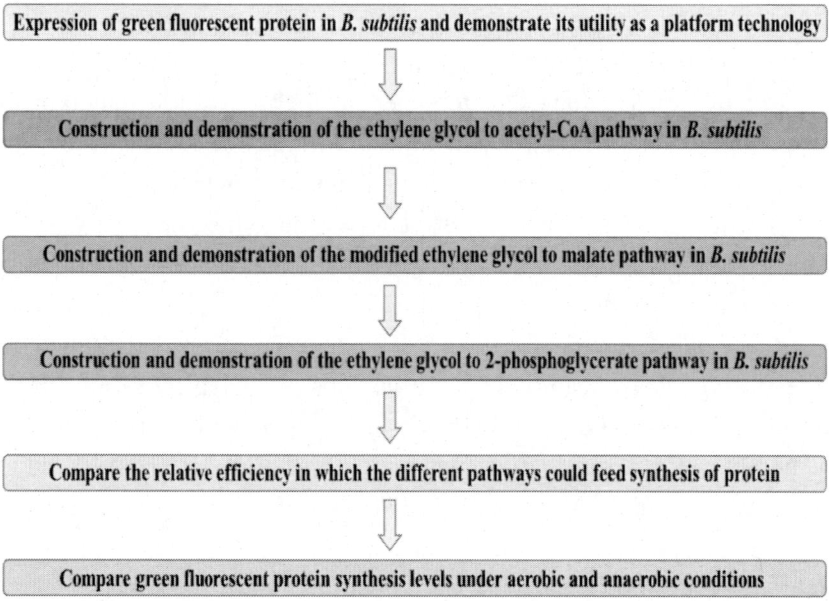

Figure C4. Schematic diagram showing the conceptual flow of the ethylene glycol to protein synthesis project.

Having established green fluorescent protein expression in *B. subtilis*, and its role as a platform technology for the study that is to be conducted in this project, experiments would next be initiated to build the ethylene glycol to acetyl-CoA pathway in the bacterium. All three genes that encode enzymes for the pathway would be encoded as an operon to be controlled by PgyrA in a plasmid vector. Initial study would examine if the pathway would help afford *B. subtilis* the ability to grow on ethylene glycol as the sole carbon source. Once this has been achieved, the next step would be the co-transformation of the plasmids encoding the pathway and green fluorescent protein into *B. subtilis* for the next phase of experimentation (Estimated time required: 4 weeks).

Next in the project is the construction of a modified ethylene glycol to malate pathway where there is no co-feeding of ethanol or acetate. Hence, all pathway genes would be encoded in a plasmid under an operon controlled by PgyrA. Similar to the first pathway, ability of *B. subtilis* to grow on ethylene glycol as sole carbon source using this new modified pathway must be assessed. If growth on ethylene glycol could be established, the next step would be to cotransform the plasmid encoding the green fluorescent protein together with the plasmid encoding the ethylene glycol utilization pathway into *B. subtilis* to establish a new strain for later comparative experimentation (Estimated time required: 4 weeks).

The last pathway to be constructed is the same one that has been explored in Future project B. Linking ethylene glycol to 2-phosphoglycerate, this pathway can be divided into 2 modular parts with glyoxylate as intermediate metabolite. Each of the pathway module would be encoded by a plasmid. In addition, one of the plasmids would be selected to encode the gene for green fluorescent protein. This is to limit the number of plasmids used to two per cell. Preliminary work would test if the dual plasmids work in *B. subtilis* to help the bacterium grow on ethylene glycol as sole carbon source with concomitant production of green fluorescent protein (Estimated time required: 3 weeks).

Completion of pathway construction and development of respective strains harbouring different ethylene glycol utilization pathway paves for their performance test to determine the best entry point in central carbon metabolism for ethylene glycol to fuel high green fluorescent protein production. Growing cells would be utilized as platform for the test at 30°C and 225 rpm, which will help determine the extent in which flux delivered to glycolytic pathway could be channelled to TCA cycle and vice versa to support amino acid synthesis and protein production (Estimated time required: 2 weeks).

Finally, the pathway constructed in this project would be examined under anaerobic cultivation conditions to assess their relative efficiency in supporting amino acid synthesis and protein production. This holds particular relevance for the ethylene glycol to malate pathway as it enters central carbon metabolism through tricarboxylic acid cycle, which may be affected by anaerobic growth conditions. Anaerobic growth experiments would be conducted at 30°C and 225 rpm with 0.5 mL of cell-growth medium suspension in 2.0 mL HPLC glass vials. Each glass vial is a sampling point to help ensure anaerobic conditions for all samples. Insights gleaned from this series of experimentations would help expand our understanding of anaerobic cell growth and its impact on production of recombinant proteins, a topic of relevance to the biotechnology industry (Estimated time required: 2 weeks).

Plasmid construction has been factored into the above timeline. Another 4 weeks would be included for general troubleshooting. Overall, the project should be completed in 23 weeks or 6 months. The timeline for the project is as listed below in Table C1.

Table C1. Timeline of experimental programme in ethylene glycol to protein synthesis project

No.	Major tasks in the project	Time required
1	Expression of green fluorescent protein in *B. subtilis* and demonstrate its utility as a platform technology	4 weeks
2	Construction and demonstration of the ethylene glycol to acetyl-CoA pathway in *B. subtilis*	4 weeks
3	Construction and demonstration of the modified ethylene glycol to malate pathway in *B. subtilis*	4 weeks
4	Construction and demonstration of the ethylene glycol to 2phosphoglycerate pathway in *B. subtilis*	3 weeks
5	Compare the relative efficiency in which the different pathways could feed synthesis of protein	2 weeks
6	Compare green fluorescent protein synthesis levels under aerobic and anaerobic conditions	2 weeks
7	General troubleshooting	4 weeks
	Total:	23 weeks

CONCLUSION

Recombinant protein synthesis is the foundation of modern biotechnology. Over the ensuing decades, much has been learned about factors that influence recombinant protein production. These factors include strength of promoter and ribosome binding sites as well as use of codon optimized genes for expressing proteins. However, relatively little is known about how entry point of substrate into central carbon metabolism influences synthesis of different amino acids, which will have an impact on overall protein production.

This work sought to use different ethylene glycol utilization pathway concepts to provide some indications on the choice of entry point for this alternative substrate for maximizing production of a marker protein, green fluorescent protein. Specifically, three different pathways that terminate at different metabolites in central carbon metabolism would be constructed and assessed for their relative efficiency in providing flux to amino acid synthesis and protein production under both aerobic and anaerobic conditions. The three terminal pathway metabolites are malate, 2-phosphoglycerate and acetyl-CoA, which enters central carbon metabolism at different entry points. This holds relevance as the intermediates that feed amino acids are situated in both glycolysis and tricarboxylic acid (TCA) cycle.

The central hypothesis tested in this work posits that proximity of entry point to metabolites that feed amino acid synthesis would have a large impact on amount of metabolic flux that could be diverted for amino acid synthesis and protein production. For example, channelling of ethylene glycol metabolic flux to 2-phosphoglycerate could only adequately fuel amino acid synthesis that feed off pyruvate, 3-phosphoglycerate and pentose phosphate pathway since they are located close to the entry point. Less flux would be diverted to oxaloacetate and α-ketoglutarate in TCA cycle, which meant that synthesis of some needed amino acids would be deficiency, which will have an impact on overall protein yield.

As designed, the research programme will start off with establishing expression of green fluorescent protein in *B. subtilis* as the top priority given its status as a platform technology for this project. Next, construction of the three different pathways will begin and strains harbouring different pathway would be developed. These strains would subsequently be used in performance tests to evaluate which entry point offers the best distribution of ethylene glycol flux for amino acid synthesis and protein production. Finally, anaerobic growth experiments would be conducted to assess if results of aerobic cell growth and protein production experiments also apply.

Overall, the proposed project will first demonstrate the utility of channelling ethylene glycol flux to acetyl-CoA in fuelling cell growth and protein production. This pathway would subsequently form the basis for comparison with two other pathways that link ethylene glycol to malate and 2-phosphoglycerate, respectively. Knowledge gained should inform our choice of entry point in *B. subtilis* central carbon metabolism for improving recombinant protein production, that hold significant implications for our understanding of utilizing the Gram-positive bacterium's metabolism for amino acid synthesis and protein production. Lessons learned from the ethylene glycol case example should be applicable to other substrates, where pathways are built to connect the substrate to different entry points in central carbon metabolism.

Chapter 6

CONCLUSION

Overall, the series of experiments conducted in the first project of the thesis validated the functionality of the ethanol utilization pathway (EUP) in enabling NADH regeneration in both single phase and two-phase biotransformation systems under anaerobic conditions in 2.0 mL HPLC vials. For example, butanone could be converted to 2-butanol in single phase biotransformation with NADH produced from ethanol conversion to acetyl-CoA. In the case of two phase biotransformation, acetophenone could be converted to 1-Phenylethanol again with the help of NADH generated by EUP pathway. Overall, the EUP pathway as implemented in *Escherichia coli*, could enable NADH regeneration when coupled with a biotransformation reaction dependent on NADH. The above coupling is tight with correlated production of transformation product (1-Phenylethanol and 2-butanol) and acetate. Aerobic and anaerobic cell growth experiments also verified that the EUP pathway could support cell growth through feeding of ethanol to acetyl-CoA.

Both resting cell and growing cell biotransformation were explored in this work to provide complementary views of the system that could inform on process performance and help identify bottlenecks that hamper, for example, biotransformation efficiency. In particular, resting cell seeks to mimic industrial conditions where cells for biotransformation are not provided with nutrients except for the biotransformation substrate and co-substrate. As construed, resting cell biotransformation is thus more cost-efficient and is less complex. On the other hand, the architecture of the ethanol utilization pathway where ethanol is converted to acetyl-CoA meant that possibility exists in using cell growth to deplete the intracellular pool of acetyl-CoA, and thereby, setting up a driving force for ethanol consumption. Hence, growing cell biotransformation are also explored in this project, and the results validated theoretical predictions. Specifically, growing cell biotransformation helped improve biotransformation efficiency compared to resting cells.

One significant finding from this series of experiment is that growing cell biotransformation works also in the case of two-phase biotransformation. Typically, the organic overlay in two-phase biotransformation is thought to induce toxicity in the cells, and thus, negatively impact on biotransformation. But, reasonably good biotransformation efficiency was obtained in two-phase acetophenone biotransformation. Ability to work in both single phase and two-phase biotransformation systems augmented EUP pathway's utility as a good candidate for NADH regeneration in both biocatalytic and metabolic engineering applications.

Substantial body of work in other fields has validated that concept of existence of differential gene expression in cells grown in different medium. In this work, two growth medium, LB and M9 + 10 g/L ethanol, were explored for their relative efficacy in enabling higher biotransformation efficiency through tuning the gene expression pattern of cells. Results revealed that cells precultivated (resting cells) or cultivated (growing cells) in M9 + 10 g/L ethanol medium persistently delivered higher biotransformation efficiency, which signifies that the gene expression pattern elicited in *E. coli* cells in M9 ethanol medium is conducive for biotransformation. This highlights an interesting area which could be further explored through the commercially available RNA-seq technology for probing the exact details of differential gene expression of cells grown in the two growth media.

To help understand the parameters that govern the biotransformation system and to validate our conceptual understanding of the process, optimization experiments were performed through tuning important parameters such as temperature, initial biotransformation substrate concentration, ethanol concentration, and promoter for EUP pathway. Specifically, temperature effect studies identified 37°C as the optimal temperature for biotransformation. Substrate concentration, on the other hand, reveals nuanced dependence that depends critically on the type of biotransformation used. Specifically, initial acetophenone concentration shows a positive correlation with biotransformation efficiency. This is in contrast to the substrate toxicity effect observed in butanone biotransformation at high substrate concentration. The divergence in system performance arises because of the protective role of the organic overlay in shielding cells from high acetophenone concentration and toxicity effect. Such protection is not available in a single phase biotransformation system such as that for butanone which is in close contact with the cells. This then explains why there is a substrate toxicity effect in butanone biotransformation system at high butanone loading.

Comparison of EUP pathway with the state-of-the-art glucose dehydrogenase system reveals that the EUP pathway is more efficient in NADH regeneration compared to glucose dehydrogenase system. This dovetails with theoretical predictions based on the structure of the two NADH regeneration pathways and amount of NADH regenerated per substrate molecule. This aspect was also indirectly validated in experiments seeking to compare the relative efficacy of endogenous glucose and ethanol utilization pathway. Results validated that endogenous ethanol utilization is more efficient in NADH regeneration and engenders higher biotransformation efficiency compared to endogenous ethanol utilization. Given that endogenous adhE has the same pathway architecture as heterologous EUP pathway, the result reaffirms ethanol utility as an efficient NADH regeneration substrate for use in biocatalysis.

Efforts to improve biotransformation efficiency saw the use of a constitutive promoter, PgyrA, in controlling the expression of EUP pathway genes. Results, however, show that the biotransformation efficiency obtained in strain with EUP under the control of PgyrA is lower than that of strain with EUP under the control of pthrC autoinducible promoter. Lower expression level of genes under PgyrA is a plausible explanation of the result and suggests that other factors may be limiting biotransformation efficiency. One possibility is the reversibility of the EUP pathway reactions that may explain why there is a fundamental constraint on ethanol utilization under anaerobic conditions. To this end, an attempt was made to modify the EUP pathway architecture by channelling acetaldehyde flux to acetate instead of acetyl-CoA. Experiment characterization of the strain with the modified EUP pathway in both resting cell and growing cell biotransformation reveals lower biotransformation efficiency compared to the original EUP pathway that ends in acetyl-CoA. Such an outcome could be due to relatively low activity of the aldA enzyme that channel flux from acetaldehyde to acetate.

Besides improving biotransformation efficiencies, efforts were also directed at exploring alternative forms of biotransformation such as mixed substrate biotransformation. In this conceptualization, glucose and ethanol would serve as the co-substrates. Ideally, glucose would power cell growth in the first phase followed by the use of ethanol in biotransformation. Hence, mixed substrate biotransformation should provide for higher biotransformation efficiency compared to what can be achieved with either glucose or ethanol alone. Results obtained, however, revealed that mixed substrate biotransformation was less efficient compared to those obtained with ethanol alone. This likely comes from the effect of glucose dependent carbon catabolite repression that reduced the expression of EUP pathway genes. Such a postulation is not without fundamental basis given that transcription of EUP pathway gene is under the control of the autoinducible pthrC promoter that may be evolutionary wired into the glucose regulon, and thus, whose transcriptional activity can be affected by the presence of glucose.

Future work in the project entails examining how the endogenous adhE enzyme compares with the two gene EUP pathway in enabling biotransformation. Given that both pathways are similar in architecture with acetyl-CoA as end-product, a one gene pathway may be preferable than the current EUP pathway in terms of reducing metabolic burden of expressing the pathway genes. Besides replacing the current EUP pathway with adhE, validating the EUP pathway at aerobic biotransformation would also expands the utility scope of the ethanol-based biotransformation system. Finally, there is a need to demonstrate the utility of EUP NADH regeneration system under aerobic biotransformation conditions to expand the application range of the system.

While glucose dehydrogenase is the dominant NADH regeneration system in biocatalysis, there remains scientific value in exploring other NADH regeneration system mediated by the utilization of other substrates. One possible substrate, ethanol, has been explored in an earlier work in this thesis proposal. Another possibility comes by accident during the formulation of a two-step utilization pathway for ethylene glycol. Both steps of the pathway generate NADH from NAD^+, and theoretically could be used for cofactor regeneration. Formation of a metabolite, glycolate, that could integrate with central carbon metabolism through glyoxylate is another factor arguing for the use of the postulated ethylene glycol utilization pathway for cofactor regeneration since it does not end with a unnatural metabolite that in accumulating in the cell, would cause toxicity problem later on. Finally, exploring the NADH regeneration capability of the ethylene glycol utilization pathway would also illuminate the relative efficiency of pathway enzymes.

The ethylene glycol utilization pathway would be combined with cpsADH to form a plasmid-based system for butanone biotransformation in *E. coli* BL21 (DE3). This system enjoys the benefit of having all metabolites, substrates and products, easily detectable by a single HPLC run given that they are all water-soluble. Preliminary findings from the project have validated the analytical concept based on HPLC detection of ethylene glycol and glycolate. Although ethylene glycol's retention time of 13.9 mins is quite close to that of acetate (13.2 mins), separation of the two analytes should not be a problem. Given that both steps of the ethylene glycol utilization pathway are powered by genes native to *E. coli*, it is important to check the endogenous capability of *E. coli* in utilizing ethylene glycol. Results show that *E. coli* could not utilize ethylene glycol to support growth, a finding which is in sync with transcriptome analysis revealing that the expression level of endogenous fucO is low. Finally, transcriptome analysis also reveals that glycolate produced by ethylene glycol utilization could be connected to glyoxylate through endogenous glyoxylate reductase.

Planned experimental work sought to first demonstrate that a heterologous ethylene glycol utilization pathway could afford *E. coli* the ability to grow on the substrate, a fact which would also show that the pathway is functional. After which, attention would switch to performing initial integration work for both parts of the biotransformation system (i.e., cofactor regeneration and biocatalytic enzyme). Substrate toxicity effect, common in single phase biotransformation system, would be the next focus given uncertainty over the toxicity of ethylene glycol at elevated concentration. Resting cell and growing cell biotransformation, and comparison of their relative efficacy at biotransformation would be a major focus of the work. Next, growth medium effect that impact on gene expression pattern would be examined through work on rich medium (LB medium) and minimal medium (M9 medium). Finally, cofeeding of acetate and ethylene glycol would be explored to improve biotransformation efficiency in growing cells.

Overall, the proposed work should help us assess if ethylene glycol utilization could help regenerate NADH in support of a biotransformation reaction in *E. coli* as well as its relative efficiency compared to glucose dehydrogenase system and ethanol utilization pathway. From the applied perspective, the work should provide insights into the system parameters that govern the functioning of the biotransformation system such as appropriate substrate concentration and whether growing cell biotransformation would enable higher biotransformation efficiency compared to resting cell biotransformation.

Need for shifting the carbon source for commodity chemicals and consumer products from fossil fuel to renewable resources has motivated the search for alternative substrates in biotechnology and metabolic engineering. Glucose has been the dominant carbon source for the generation of a variety of bioproducts over the past few decades. This comes about due to the status of glucose as the preferred carbon source for many microbial species across the tree of life, which meant that metabolic pathways are available in most species for its metabolization. Such a case is not common for many of the alternative substrates that have received some attention for pathway engineering in the past decades. Specifically, a new pathway needs to be constructed using existing enzymes from different species or entirely new enzymes to mediate the metabolization of a new alternative substrate and connect the carbon flux to central carbon metabolism.

One possible alternative substrate is ethylene glycol that could be derived from hydrolysis of lignocellulosic waste or plastic PET waste. Adequate supply of reasonably priced ethylene glycol should be available under the above scenario. But, what is more fundamental about this project is the demonstration of the feasibility of building a metabolic route from ethylene glycol to central carbon metabolism in *Bacillus subtilis* that is devoid of the glyoxylate shunt and glycerate pathways. Such a microbial chassis places an inevitable challenge on the project, but the disadvantages inherent in this challenge is modulated by the Generally regarded as safe status of *B. subtilis*.

Besides building a feasible pathway that could bring carbon flux from ethylene glycol to central carbon metabolism for building biomass, another major hypothesis tested in this project is whether different entry points in central carbon metabolism presents different efficiency in generation of biomass precursors for cell growth. Hence, two routes are planned for bringing carbon flux from ethylene glycol to central carbon metabolism. The first path connects ethylene glycol to malate in tricarboxylic acid cycle through glyoxylate, which serves as an important metabolic branch point. To help condense glyoxylate to acetyl-CoA for forming malate, an exogenous source of carbon must be supplied to provision acetyl-CoA for the purpose. Ethanol coupled with the ethanol utilization pathway explored in first project of this thesis is a viable pathway for providing acetyl-CoA. Alternatively, acetate could also be converted into acetyl-CoA through endogenous acetyl CoA synthetase in *B. subtilis*. As construed, co-feeding of ethylene glycol and ethanol or acetate is required to connect ethylene glycol to malate. Since ethanol or acetate may have different efficacy in generating acetyl-CoA for condensing glyoxylate, both substrates would be compared to determine the preferable route for generating biomass from ethylene glycol.

The second pathway involves the sole feeding of ethylene glycol and concerns the connection of the substrate to 2-phosphoglycerate in glycolysis. Similar to the ethylene glycol to malate route, glyoxylate is the metabolic branch point that divides the first half of the pathway from ethylene glycol to glyoxylate from the second half that runs to 2phosphoglycerate. The two halves of the pathway represent distinct pathway module that could be individually encoded in separate plasmids; thereby, allowing the expression level of the pathway to be modulated to help maintain a constant flux along the entire pathway. Such modulation is necessary given that activities of the enzymes in separate pathway module may be different.

Both pathways for connecting ethylene glycol to central carbon metabolism would be tested for biomass formation under aerobic conditions. But, given that some biotechnology applications require anaerobic growth, there is practical interest in understanding the efficiency of both pathways to generate biomass under anaerobic conditions. Such applied experimentation may also yield insights in basic biology on utilization of alternative substrate under different oxygen environment. From the theoretical perspective, anaerobic conditions are expected to have an impact on the ethylene glycol to malate route given that tricarboxylic acid cycle may not function optimally under these conditions. Ethylene glycol to 2phosphoglycerate route, on the other hand, should be invariant under different oxygen tension. Finally, ^{13}C carbon labelling experiments would be conducted to track and confirm the directionality of the metabolic flux flow from ethylene glycol to central carbon metabolism. Use of whole-cell lysate and ion-exchange liquid chromatography-mass spectrometry should provide a glimpse of labelling of pathway intermediates such as glycolate, glyoxylate, malate or pyruvate for confirming that the constructed pathway has indeed channelled flux to the designated connecting molecule in central carbon metabolism.

Overall, in constructing pathway connecting ethylene glycol to central carbon metabolism using genes cloned from *E. coli*, the study hopes to demonstrate the feasibility of wiring an ethylene glycol naïve *B. subtilis* for use in biotechnology revolving tapping the carbon flux from the alternative C2 substrate. Challenges involved in this endeavour include the expression of a number of heterologous pathway genes in a host that may succumb to excessive metabolic burden, which meant that modulation of expression level is critical for experiment success. Comparison between connecting ethylene glycol to glycolysis or tricarboxylic acid cycle would provide important data indicating the relative efficiency in which glycolysis or tricarboxylic acid cycle in channelling flux for building biomass. This constitute important data that could guide our understanding in selecting appropriate entry points for connecting flux from alternative substrates. Such a choice is likely species-specific given differing relative abundances of different enzymes in central carbon metabolism in different species. But, its core essence is similar to the evolutionary forced selection of particular entry points in central carbon metabolism for the commonly used substrates such as glycerol, fructose and glucose.

Recombinant protein synthesis is the foundation of modern biotechnology. Over the ensuing decades, much has been learned about factors that influence recombinant protein production. These factors include strength of promoter and ribosome binding sites as well as use of codon optimized genes for expressing proteins. However, relatively little is known about how entry point of substrate into central carbon metabolism influences synthesis of different amino acids, which will have an impact on overall protein production.

This work sought to use different ethylene glycol utilization pathway concepts to provide some indications on the choice of entry point for this alternative substrate for maximizing production of a marker protein, green fluorescent protein. Specifically, three different pathways that terminate at different metabolites in central carbon metabolism would be constructed and assessed for their relative efficiency in providing flux to amino acid synthesis and protein production under both aerobic and anaerobic conditions. The three terminal pathway metabolites are malate, 2-

phosphoglycerate and acetyl-CoA, which enters central carbon metabolism at different entry points. This holds relevance as the intermediates that feed amino acids are situated in both glycolysis and tricarboxylic acid (TCA) cycle.

The central hypothesis tested in this work posits that proximity of entry point to metabolites that feed amino acid synthesis would have a large impact on amount of metabolic flux that could be diverted for amino acid synthesis and protein production. For example, channelling of ethylene glycol metabolic flux to 2-phosphoglycerate could only adequately fuel amino acid synthesis that feed off pyruvate, 3-phosphoglycerate and pentose phosphate pathway since they are located close to the entry point. Less flux would be diverted to oxaloacetate and α-ketoglutarate in TCA cycle, which meant that synthesis of some needed amino acids would be deficiency, which will have an impact on overall protein yield.

As designed, the research programme will start off with establishing expression of green fluorescent protein in *B. subtilis* as the top priority given its status as a platform technology for this project. Next, construction of the three different pathways will begin and strains harbouring different pathway would be developed. These strains would subsequently be used in performance tests to evaluate which entry point offers the best distribution of ethylene glycol flux for amino acid synthesis and protein production. Finally, anaerobic growth experiments would be conducted to assess if results of aerobic cell growth and protein production experiments also apply.

Overall, the proposed project will first demonstrate the utility of channelling ethylene glycol flux to acetyl-CoA in fuelling cell growth and protein production. This pathway would subsequently form the basis for comparison with two other pathways that link ethylene glycol to malate and 2-phosphoglycerate, respectively. Knowledge gained should inform our choice of entry point in *B. subtilis* central carbon metabolism for improving recombinant protein production, that hold significant implications for our understanding of utilizing the Gram-positive bacterium's metabolism for amino acid synthesis and protein production. Lessons learned from the ethylene glycol case example should be applicable to other substrates, where pathways are built to connect the substrate to different entry points in central carbon metabolism.

REFERENCES

[1] Ledesma-Amaro, R. & Nicaud, J. M. Metabolic Engineering for Expanding the Substrate Range of Yarrowia lipolytica. *Trends in Biotechnology* 34, 798-809, doi:https://doi.org/10.1016/j.tibtech.2016.04.010 (2016).

[2] Buschke, N. et al. Systems metabolic engineering of xylose-utilizing Corynebacterium glutamicum for production of 1,5-diaminopentane. *Biotechnology Journal* 8, 557-570, doi:10.1002/biot.201200367 (2013).

[3] Lazar, Z., Dulermo, T., Neuvéglise, C., Crutz-Le Coq, A. M. & Nicaud, J. M. Hexokinase—A limiting factor in lipid production from fructose in Yarrowia lipolytica. *Metabolic Engineering* 26, 89-99, doi:https://doi.org/10.1016/j.ymben.2014.09.008 (2014).

[4] Blankschien, M. D., Clomburg, J. M. & Gonzalez, R. Metabolic engineering of *Escherichia coli* for the production of succinate from glycerol. *Metabolic Engineering* 12, 409-419, doi:https://doi.org/10.1016/j.ymben.2010.06.002 (2010).

[5] van de Laar, T. et al. Increased heterologous protein production by Saccharomyces cerevisiae growing on ethanol as sole carbon source. *Biotechnology and Bioengineering* 96, 483-494, doi:10.1002/bit.21150 (2007).

[6] Yuan, Z. et al. High production of pyoluteorin and 2,4-diacetylphloroglucinol by Pseudomonas fluorescens S272 grown on ethanol as a sole carbon source. *Journal of Fermentation and Bioengineering* 86, 559-563, doi:https://doi.org/10.1016/S0922338X(99)80006-3 (1998).

[7] Membrillo-Hernández, J. et al. Evolution of the adhE Gene Product of *Escherichia coli* from a Functional Reductase to a Dehydrogenase: Genetic and Biochemical Studies of the Mutant Proteins. *Journal of Biological Chemistry* 275, 33869-33875 (2000).

[8] Boronat, A., Caballero, E. & Aguilar, J. Experimental evolution of a metabolic pathway for ethylene glycol utilization by *Escherichia coli*. *Journal of Bacteriology* 153, 134 (1983).

[9] Brockmeier, U. et al. Systematic Screening of All Signal Peptides from Bacillus subtilis: A Powerful Strategy in Optimizing Heterologous Protein Secretion in Grampositive Bacteria. *Journal of Molecular Biology* 362, 393-402, doi:https://doi.org/10.1016/j.jmb.2006.07.034 (2006).

[10] Westers, L., Westers, H. & Quax, W. J. Bacillus subtilis as cell factory for pharmaceutical proteins: a biotechnological approach to optimize the host organism. *Biochimica et Biophysica Acta (BBA) - Molecular Cell Research* 1694, 299-310, doi:https://doi.org/10.1016/j.bbamcr.2004.02.011 (2004).

[11] Yun, H., Choi, H. L., Fadnavis, N. W. & Kim, B. G. Stereospecific Synthesis of (R)2-Hydroxy Carboxylic Acids Using Recombinant *E. coli* BL21 Overexpressing YiaE from *Escherichia coli* K12 and Glucose Dehydrogenase from Bacillus subtilis. *Biotechnology Progress* 21, 366-371, doi:10.1021/bp049694w (2005).

[12] Wang, Y., San, K. Y. & Bennett, G. N. Cofactor engineering for advancing chemical biotechnology. *Current Opinion in Biotechnology* 24, 994-999, doi:https://doi.org/10.1016/j.copbio.2013.03.022 (2013).

[13] Mouri, T., Shimizu, T., Kamiya, N., Goto, M. & Ichinose, H. Design of a cytochrome P450BM3 reaction system linked by two-step

cofactor regeneration catalyzed by a soluble transhydrogenase and glycerol dehydrogenase. *Biotechnology Progress* 25, 1372-1378, doi:10.1002/btpr.231 (2009).

[14] Mouri, T., Michizoe, J., Ichinose, H., Kamiya, N. & Goto, M. A recombinant *Escherichia coli* whole cell biocatalyst harboring a cytochrome P450cam monooxygenase system coupled with enzymatic cofactor regeneration. *Applied Microbiology and Biotechnology* 72, 514-520, doi:10.1007/s00253-005-0289-y (2006).

[15] Bäumchen, C. et al. D-Mannitol production by resting state whole cell biotransformation of D-fructose by heterologous mannitol and formate dehydrogenase gene expression in Bacillus megaterium. *Biotechnology Journal* 2, 1408-1416, doi:10.1002/biot.200700055 (2007).

[16] Riebel, B. R., Gibbs, P. R., Wellborn, W. B. & Bommarius, A. S. Cofactor Regeneration of NAD+ from NADH: Novel Water-Forming NADH Oxidases. *Advanced Synthesis & Catalysis* 344, 1156-1168, doi:10.1002/16154169(200212)344:10<1156::AID-ADSC1156>3.0.CO;2-# (2002).

[17] Riebel, B. R., Gibbs, P. R., Wellborn, W. B. & Bommarius, A. S. Cofactor Regeneration of both NAD+ from NADH and NADP+ from NADPH:NADH Oxidase from Lactobacillus sanfranciscensis. *Advanced Synthesis & Catalysis* 345, 707-712, doi:10.1002/adsc.200303039 (2003).

[18] Ye, Q. et al. Biosynthesis of (S)-4-chloro-3-hydroxybutanoate ethyl using *Escherichia coli* co-expressing a novel NADH-dependent carbonyl reductase and a glucose dehydrogenase. *Bioresource Technology* 101, 8911-8914, doi:https://doi.org/10.1016/j.biortcch.2010.06.098 (2010).

[19] Xu, Z., Jing, K., Liu, Y. & Cen, P. High-level expression of recombinant glucose dehydrogenase and its application in NADPH regeneration. *Journal of Industrial Microbiology & Biotechnology* 34, 83-90, doi:10.1007/s10295 006-0168-2 (2007).

[20] Pongtharangkul, T. et al. Kinetic properties and stability of glucose dehydrogenase from Bacillus amyloliquefaciens SB5 and its potential for cofactor regeneration. *AMB Express* 5, 68, doi:10.1186/s13568-015-0157-9 (2015).

[21] Wong, C. H. & Whitesides, G. M. Enzyme-catalyzed organic synthesis: NAD(P)H cofactor regeneration using ethanol/alcohol dehydrogenase/aldehyde dehydrogenase and methanol/alcohol dehydrogenase/aldehyde dehydrogenase/formate dehydrogenase. *The Journal of Organic Chemistry* 47, 2816-2818, doi:10.1021/jo00135a037 (1982).

[22] Broussy, S., Cheloha, R. W. & Berkowitz, D. B. Enantioselective, KetoreductaseBased Entry into Pharmaceutical Building Blocks: Ethanol as Tunable Nicotinamide Reductant. *Organic Letters* 11, 305-308, doi:10.1021/ol802464g (2009).

[23] Schroer, K., Peter Luef, K., Stefan Hartner, F., Glieder, A. & Pscheidt, B. Engineering the Pichia pastoris methanol oxidation pathway for improved NADH regeneration during whole-cell biotransformation. *Metabolic Engineering* 12, 8-17, doi:https://doi.org/10.1016/j.ymben.2009.08.006 (2010).

[24] Wang, X. et al. Cofactor NAD(P)H Regeneration Inspired by Heterogeneous Pathways. *Chem* 2, 621-654, doi:https://doi.org/10.1016/j.chempr.2017.04.009 (2017).

[25] Kim, S. J. et al. Enhanced production of 2,3-butanediol from xylose by combinatorial engineering of xylose metabolic pathway and cofactor regeneration in pyruvate decarboxylase-deficient Saccharomyces cerevisiae. *Bioresource Technology* 245, 1551-1557, doi:https://doi.org/10.1016/j.biortech.2017.06.034 (2017).

[26] Peters, J., Minuth, T. & Kula, M. R. A novel NADH-dependent carbonyl reductase with an extremely broad substrate range from Candida parapsilosis: Purification and characterization. *Enzyme and Microbial Technology* 15, 950-958, doi:https://doi.org/10.1016/0141-0229(93)90171-W (1993).

[27] Jakoblinnert, A. et al. Design of an activity and stability improved carbonyl reductase from Candida parapsilosis. *Journal of*

Biotechnology 165, 52-62, doi:https://doi.org/10.1016/j.jbiotec.2013.02.006 (2013).

[28] Peters, J., Minuth, T. & Kula, M. R. Kinetic and Mechanistic Studies of a Novel Carbonyl Reductase Isolated from Candida Parapsilosis. *Biocatalysis* 8, 31-46, doi:10.3109/10242429309030954 (1993).

[29] Ni, Y. & Xu, J. H. Biocatalytic ketone reduction: A green and efficient access to enantiopure alcohols. *Biotechnology Advances* 30, 1279-1288, doi:https://doi.org/10.1016/j.biotechadv.2011.10.007 (2012).

[30] Valadez-Blanco, R. & Livingston, A. G. Enantioselective whole-cell biotransformation of acetophenone to S-phenylethanol by Rhodotorula glutinis: Part I. Product formation kinetics and feeding strategies in aqueous media. *Biochemical Engineering Journal* 46, 44-53, doi:https://doi.org/10.1016/j.bej.2009.05.009 (2009).

[31] Erdmann, V., Mackfeld, U., Rother, D. & Jakoblinnert, A. Enantioselective, continuous (R)- and (S)-2-butanol synthesis: Achieving high space-time yields with recombinant *E. coli* cells in a micro-aqueous, solvent-free reaction system. *Journal of Biotechnology* 191, 106-112, doi:https://doi.org/10.1016/j.jbiotec.2014.06.032 (2014).

[32] Wachtmeister, J. & Rother, D. Recent advances in whole cell biocatalysis techniques bridging from investigative to industrial scale. *Current Opinion in Biotechnology* 42, 169-177, doi:https://doi.org/10.1016/j.copbio.2016.05.005 (2016).

[33] Jeon, E. Y. et al. Intracellular transformation rates of fatty acids are influenced by expression of the fatty acid transporter FadL in *Escherichia coli* cell membrane. *Journal of Biotechnology* 281, 161-167, doi:https://doi.org/10.1016/j.jbiotec.2018.07.019 (2018).

[34] Zhang, C., Chen, X., Stephanopoulos, G. & Too, H. P. Efflux transporter engineering markedly improves amorphadiene production in *Escherichia coli*. *Biotechnology and Bioengineering* 113, 1755-1763, doi:10.1002/bit.25943 (2016).

[35] Li, X., Jiang, B. & Pan, B. Biotransformation of phenylpyruvic acid to phenyllactic acid by growing and resting cells of a Lactobacillus

sp. *Biotechnology Letters* 29, 593-597, doi:10.1007/s10529-006-9275-4 (2007).

[36] Zhang, Y. et al. The two-step biotransformation of monosodium glutamate to GABA by Lactobacillus brevis growing and resting cells. *Applied Microbiology and Biotechnology* 94, 1619-1627, doi:10.1007/s00253-012-3868-8 (2012).

[37] Ma, X. et al. A standard for near-scarless plasmid construction using reusable DNA parts. *Nature Communications* 10, 3294, doi:10.1038/s41467-019-11263-0 (2019).

[38] Soufi, B., Krug, K., Harst, A. & Macek, B. Characterization of the *E. coli* proteome and its modifications during growth and ethanol stress. *Frontiers in Microbiology* 6, 103 (2015).

[39] Haft, R. J. F. et al. Correcting direct effects of ethanol on translation and transcription machinery confers ethanol tolerance in bacteria. *Proceedings of the National Academy of Sciences* 111, E2576, doi:10.1073/pnas.1401853111 (2014).

[40] Yomano, L. P., York, S. W. & Ingram, L. O. Isolation and characterization of ethanoltolerant mutants of *Escherichia coli* KO11 for fuel ethanol production. *Journal of Industrial Microbiology and Biotechnology* 20, 132-138, doi:10.1038/sj.jim.2900496 (1998).

[41] Wang, Y. et al. Adaptive evolution of nontransgenic *Escherichia coli* KC01 for improved ethanol tolerance and homoethanol fermentation from xylose. *Journal of Industrial Microbiology & Biotechnology* 38, 1371-1377, doi:10.1007/s10295-0100920-5 (2011).

[42] Goodarzi, H. et al. Regulatory and metabolic rewiring during laboratory evolution of ethanol tolerance in *E. coli*. *Molecular Systems Biology* 6, 378, doi:10.1038/msb.2010.33 (2010).

[43] Horinouchi, T. et al. Transcriptome analysis of parallel-evolved *Escherichia coli* strains under ethanol stress. *BMC Genomics* 11, 579, doi:10.1186/1471-2164-11-579 (2010).

[44] Woodruff, L. B. A. et al. Genome-scale identification and characterization of ethanol tolerance genes in *Escherichia coli*.

Metabolic Engineering 15, 124-133, doi:https://doi.org/10.1016/j.ymben.2012.10.007 (2013).

[45] Dombek, K. M. & Ingram, L. O. Effects of ethanol on the *Escherichia coli* plasma membrane. *Journal of Bacteriology* 157, 233 (1984).

[46] Liu, W. T. et al. Imaging mass spectrometry of intraspecies metabolic exchange revealed the cannibalistic factors of Bacillus subtilis. *Proceedings of the National Academy of Sciences* 107, 16286, doi:10.1073/pnas.1008368107 (2010).

[47] Lamsa, A., Liu, W. T., Dorrestein, P. C. & Pogliano, K. The Bacillus subtilis cannibalism toxin SDP collapses the proton motive force and induces autolysis. *Molecular Microbiology* 84, 486-500, doi:10.1111/j.1365-2958.2012.08038.x (2012).

[48] Chen, X., Li, S. & Liu, L. Engineering redox balance through cofactor systems. *Trends in Biotechnology* 32, 337-343, doi:https://doi.org/10.1016/j.tibtech.2014.04.003 (2014).

[49] Heux, S., Cachon, R. & Dequin, S. Cofactor engineering in Saccharomyces cerevisiae: Expression of a H2O-forming NADH oxidase and impact on redox metabolism. *Metabolic Engineering* 8, 303-314, doi:https://doi.org/10.1016/j.ymben.2005.12.003 (2006).

[50] Singh, A., Lynch, M. D. & Gill, R. T. Genes restoring redox balance in fermentationdeficient *E. coli* NZN111. *Metabolic Engineering* 11, 347-354, doi:https://doi.org/10.1016/j.ymben.2009.07.002 (2009).

[51] Liu, W. & Wang, P. Cofactor regeneration for sustainable enzymatic biosynthesis. *Biotechnology Advances* 25, 369-384, doi:https://doi.org/10.1016/j.biotechadv.2007.03.002 (2007).

[52] Johannes, T. W., Woodyer, R. D. & Zhao, H. Efficient regeneration of NADPH using an engineered phosphite dehydrogenase. *Biotechnology and Bioengineering* 96, 1826, doi:10.1002/bit.21168 (2007).

[53] Lee, W. H., Kim, M. D., Jin, Y. S. & Seo, J. H. Engineering of NADPH regenerators in *Escherichia coli* for enhanced

biotransformation. *Applied Microbiology and Biotechnology* 97, 2761-2772, doi:10.1007/s00253-013-4750-z (2013).

[54] Kim, D.-M. & Swartz, J. R. Prolonging cell-free protein synthesis with a novel ATP regeneration system. *Biotechnology and Bioengineering* 66, 180-188, doi:10.1002/(SICI)1097-0290(1999)66:3<180::AID-BIT6>3.0.CO;2-S (1999).

[55] Resnick, S. M. & Zehnder, A. J. B. In Vitro ATP Regeneration from Polyphosphate and AMP by Polyphosphate: AMP Phosphotransferase and Adenylate Kinase from <em>Acinetobacter johnsonii</em> 210A. *Applied and Environmental Microbiology* 66, 2045, doi:10.1128/AEM.66.5.2045-2051.2000 (2000).

[56] Kara, S., Schrittwieser, J. H., Hollmann, F. & Ansorge-Schumacher, M. B. Recent trends and novel concepts in cofactor-dependent biotransformations. *Applied Microbiology and Biotechnology* 98, 1517-1529, doi:10.1007/s00253-013-5441-5 (2014).

[57] Ng, C. Y., Farasat, I., Maranas, C. D. & Salis, H. M. Rational design of a synthetic Entner–Doudoroff pathway for improved and controllable NADPH regeneration. *Metabolic Engineering* 29, 86-96, doi:https://doi.org/10.1016/j.ymben.2015.03.001 (2015).

[58] Zhang, X., Zhang, B., Lin, J. & Wei, D. Oxidation of ethylene glycol to glycolaldehyde using a highly selective alcohol dehydrogenase from Gluconobacter oxydans. *Journal of Molecular Catalysis B: Enzymatic* 112, 69-75, doi:https://doi.org/10.1016/j.molcatb.2014.12.006 (2015).

[59] Ichinose, H., Kamiya, N. & Goto, M. Enzymatic Redox Cofactor Regeneration in Organic Media: Functionalization and Application of Glycerol Dehydrogenase and Soluble Transhydrogenase in Reverse Micelles. *Biotechnology Progress* 21, 11921197, doi:10.1021/bp0500765 (2005).

[60] Nowak, C., Pick, A., Lommes, P. & Sieber, V. Enzymatic Reduction of Nicotinamide Biomimetic Cofactors Using an Engineered Glucose Dehydrogenase: Providing a Regeneration System for

Artificial Cofactors. *ACS Catalysis* 7, 5202-5208, doi:10.1021/acscatal.7b00721 (2017).

[61] Zhang, K., Sawaya, M. R., Eisenberg, D. S. & Liao, J. C. Expanding metabolism for biosynthesis of nonnatural alcohols. *Proceedings of the National Academy of Sciences* 105, 20653, doi:10.1073/pnas.0807157106 (2008).

[62] Tai, Y. S. & Zhang, K. C1 metabolism redesigned. *Nature Chemical Biology* 11, 384386, doi:10.1038/nchembio.1819 (2015).

[63] Yishai, O., Goldbach, L., Tenenboim, H., Lindner, S. N. & Bar-Even, A. Engineered Assimilation of Exogenous and Endogenous Formate in *Escherichia coli*. *ACS Synthetic Biology* 6, 1722-1731, doi:10.1021/acssynbio.7b00086 (2017).

[64] Siegel, J. B. et al. Computational protein design enables a novel one-carbon assimilation pathway. *Proceedings of the National Academy of Sciences* 112, 3704, doi:10.1073/pnas.1500545112 (2015).

[65] Yishai, O., Lindner, S. N., Gonzalez de la Cruz, J., Tenenboim, H. & Bar-Even, A. The formate bio-economy. *Current Opinion in Chemical Biology* 35, 1-9, doi:https://doi.org/10.1016/j.cbpa.2016.07.005 (2016).

[66] Bar-Even, A. Formate Assimilation: The Metabolic Architecture of Natural and Synthetic Pathways. *Biochemistry* 55, 3851-3863, doi:10.1021/acs.biochem.6b00495 (2016).

[67] Boyle, N. R. & Morgan, J. A. Computation of metabolic fluxes and efficiencies for biological carbon dioxide fixation. *Metabolic Engineering* 13, 150-158, doi:https://doi.org/10.1016/j.ymben.2011.01.005 (2011).

[68] Hawkins, A. S., McTernan, P. M., Lian, H., Kelly, R. M. & Adams, M. W. W. Biological conversion of carbon dioxide and hydrogen into liquid fuels and industrial chemicals. *Current Opinion in Biotechnology* 24, 376-384, doi:https://doi.org/10.1016/j.copbio.2013.02.017 (2013).

[69] Guadalupe-Medina, V. et al. Carbon dioxide fixation by Calvin-Cycle enzymes improves ethanol yield in yeast. *Biotechnology for Biofuels* 6, 125, doi:10.1186/17546834-6-125 (2013).

[70] Patzek, T. W. Thermodynamics of the Corn-Ethanol Biofuel Cycle. *Critical Reviews in Plant Sciences* 23, 519-567, doi:10.1080/07352680490886905 (2004).

[71] Kabisch, J. et al. Metabolic engineering of Bacillus subtilis for growth on overflow metabolites. *Microbial Cell Factories* 12, 72, doi:10.1186/1475-2859-12-72 (2013).

[72] Yoshida, S. et al. A bacterium that degrades and assimilates poly(ethylene terephthalate). *Science* 351, 1196, doi:10.1126/ science.aad6359 (2016).

[73] Gaston, L. W. & Stadtman, E. R. Fermentation of ethylene glycol by Clostridium glycolicum sp. *Journal of Bacteriology* 85, 356 (1963).

[74] Dwyer, D. F. & Tiedje, J. M. Degradation of ethylene glycol and polyethylene glycols by methanogenic consortia. *Applied and Environmental Microbiology* 46, 185 (1983).

[75] Carnegie, D. & Ramsay, J. A. Anaerobic ethylene glycol degradation by microorganisms in poplar and willow rhizospheres. *Biodegradation* 20, 551-558, doi:10.1007/s10532-008-9244-9 (2009).

[76] Colucci, M. S. & Inniss, W. E. Ethylene Glycol Utilization, Cold and Ethylene Glycol Shockand Acclimation Proteins in a Psychrotrophic Bacterium. *Current Microbiology* 32, 179-182, doi:10.1007/s002849900032 (1996).

[77] Salusjärvi, L., Havukainen, S., Koivistoinen, O. & Toivari, M. Biotechnological production of glycolic acid and ethylene glycol: current state and perspectives. *Applied Microbiology and Biotechnology* 103, 2525-2535, doi:10.1007/s00253-01909640-2 (2019).

[78] Liu, H. et al. Biosynthesis of ethylene glycol in *Escherichia coli*. *Applied Microbiology and Biotechnology* 97, 3409-3417, doi:10.1007/s00253-012-4618-7 (2013).

[79] Wei, G. et al. High cell density fermentation of Gluconobacter oxydans DSM 2003 for glycolic acid production. *Journal of*

Industrial Microbiology & Biotechnology 36, 1029-1034, doi:10.1007/s10295-009-0584-1 (2009).

[80] Wiegant, W. M. & De Bont, J. A. M. A New Route for Ethylene Glycol Metabolism in Mycobacterium E44. *Microbiology* 120, 325-331, doi:https://doi.org/10.1099/00221287-120-2-325 (1980).

[81] Franden, M. A. et al. Engineering Pseudomonas putida KT2440 for efficient ethylene glycol utilization. *Metabolic Engineering* 48, 197-207, doi:https://doi.org/10.1016/j.ymben.2018.06.003 (2018).

[82] Engelberg-Kulka, H. & Hazan, R. Cannibals Defy Starvation and Avoid Sporulation. *Science* 301, 467, doi:10.1126/science.1088051 (2003).

[83] Nandy, S. K., Prasad, V. & Venkatesh, K. V. Effect of Temperature on the Cannibalistic Behavior of Bacillus subtilis. *Applied and Environmental Microbiology* 74, 7427, doi:10.1128/AEM.00683-08 (2008).

[84] Bohnenkamp, A. C. et al. Multilevel optimisation of anaerobic ethyl acetate production in engineered *Escherichia coli*. *Biotechnology for Biofuels* 13, 65, doi:10.1186/s13068-020-01703-1 (2020).

[85] Williams, D. C., Van Frank, R. M., Muth, W. L. & Burnett, J. P. Cytoplasmic inclusion bodies in *Escherichia coli* producing biosynthetic human insulin proteins. *Science* 215, 687, doi:10.1126/science.7036343 (1982).

[86] Krivoruchko, A., Zhang, Y., Siewers, V., Chen, Y. & Nielsen, J. Microbial acetylCoA metabolism and metabolic engineering. *Metabolic Engineering* 28, 28-42, doi:https://doi.org/10.1016/j.ymben.2014.11.009 (2015).

[87] Chen, Y., Daviet, L., Schalk, M., Siewers, V. & Nielsen, J. Establishing a platform cell factory through engineering of yeast acetyl-CoA metabolism. *Metabolic Engineering* 15, 48-54, doi:https://doi.org/10.1016/j.ymben.2012.11.002 (2013).

[88] Vadali, R. V., Bennett, G. N. & San, K. Y. Cofactor engineering of intracellular CoA/acetyl-CoA and its effect on metabolic flux redistribution in *Escherichia coli*. *Metabolic Engineering* 6, 133-139, doi:https://doi.org/10.1016/j.ymben.2004.02.001 (2004).

[89] Calhoun, K. A. & Swartz, J. R. Energizing cell-free protein synthesis with glucose metabolism. *Biotechnology and Bioengineering* 90, 606-613, doi:10.1002/bit.20449 (2005).

[90] Lu, X. et al. Constructing a synthetic pathway for acetyl-coenzyme A from onecarbon through enzyme design. *Nature Communications* 10, 1378, doi:10.1038/s41467-019-09095-z (2019).

[91] Wong, S. L. Advances in the use of Bacillus subtilis for the expression and secretion of heterologous proteins. *Current Opinion in Biotechnology* 6, 517-522, doi:https://doi.org/10.1016/0958-1669(95)80085-9 (1995).

[92] Ye, R. et al. High-level secretory production of intact, biologically active staphylokinase from Bacillus subtilis. *Biotechnology and Bioengineering* 62, 87-96, doi:10.1002/(SICI)1097-0290(19990105)62:1<87::AID-BIT10>3.0.CO;2-I (1999).

[93] Makino, T., Skretas, G. & Georgiou, G. Strain engineering for improved expression of recombinant proteins in bacteria. *Microbial Cell Factories* 10, 32, doi:10.1186/1475-2859-10-32 (2011).

[94] Gupta, S. K. & Shukla, P. Advanced technologies for improved expression of recombinant proteins in bacteria: perspectives and applications. *Critical Reviews in Biotechnology* 36, 1089-1098, doi:10.3109/07388551.2015.1084264 (2016).

[95] Mahalik, S., Sharma, A. K. & Mukherjee, K. J. Genome engineering for improved recombinant protein expression in *Escherichia coli. Microbial Cell Factories* 13, 177, doi:10.1186/s12934-014-0177-1 (2014).

[96] Kopp, J. et al. Boosting Recombinant Inclusion Body Production—From Classical Fed-Batch Approach to Continuous Cultivation. *Frontiers in Bioengineering and Biotechnology* 7, 297 (2019).

[97] Schumann, W. in *Advances in Applied Microbiology* Vol. 62 137-189 (Academic Press, 2007).

[98] Tripathi, N. K. & Shrivastava, A. Recent Developments in Bioprocessing of Recombinant Proteins: Expression Hosts and Process Development. *Frontiers in Bioengineering and Biotechnology* 7, 420 (2019).

[99] Vestergaard, M., Chan, S. H. J. & Jensen, P. R. Can microbes compete with cows for sustainable protein production - A feasibility study on high quality protein. *Scientific Reports* 6, 36421, doi:10.1038/srep36421 (2016).

[100] Zhang, K., Su, L. & Wu, J. Recent Advances in Recombinant Protein Production by Bacillus subtilis. *Annual Review of Food Science and Technology* 11, 295-318, doi:10.1146/annurev-food-032519-051750 (2020).

[101] Cao, H., Villatoro-Hernandez, J., Weme, R. D. O., Frenzel, E. & Kuipers, O. P. Boosting heterologous protein production yield by adjusting global nitrogen and carbon metabolic regulatory networks in Bacillus subtilis. *Metabolic Engineering* 49, 143-152, doi:https://doi.org/10.1016/j.ymben.2018.08.001 (2018).

[102] van Dijl, J. & Hecker, M. Bacillus subtilis: from soil bacterium to super-secreting cell factory. *Microbial Cell Factories* 12, 3, doi:10.1186/1475-2859-12-3 (2013).

[103] Zhang, K., Su, L., Duan, X., Liu, L. & Wu, J. High-level extracellular protein production in Bacillus subtilis using an optimized dual-promoter expression system. *Microbial Cell Factories* 16, 32, doi:10.1186/s12934-017-0649-1 (2017).

[104] Mullineaux, C. W. Classic Spotlight: Green Fluorescent Protein in <span class="named-content genus-species" id="named-content-1">Bacillussubtilis</span> and the Birth of Bacterial Cell Biology. *Journal of Bacteriology* 198, 2141, doi:10.1128/JB.00380-16 (2016).

[105] Overkamp, W. et al. Benchmarking Various Green Fluorescent Protein Variants in <span class="named-content genus-species" id="named-content1">Bacillus subtilis</span>, <span class="named-content genus-species" id="named-content-2">Strepto-coccus pneumoniae</span>, and <span class="named-content genusspecies" id="named-content-3">Lactococcus lactis</span> for Live Cell

Imaging. *Applied and Environmental Microbiology* 79, 6481, doi:10.1128/AEM.02033-13 (2013).

[106] Cao, H. & Kuipers, O. P. Influence of global gene regulatory networks on single cell heterogeneity of green fluorescent protein production in Bacillus subtilis. *Microbial Cell Factories* 17, 134, doi:10.1186/s12934-018-0985-9 (2018).

[107] Papagianni, M. Recent advances in engineering the central carbon metabolism of industrially important bacteria. *Microbial Cell Factories* 11, 50, doi:10.1186/14752859-11-50 (2012).

[108] Dauner, M., Storni, T. & Sauer, U. Bacillus subtilis Metabolism and Energetics in Carbon-Limited and Excess-Carbon Chemostat Culture. *Journal of Bacteriology* 183, 7308, doi:10.1128/JB.183.24.7308-7317.2001 (2001).

[109] Massaiu, I. et al. Integration of enzymatic data in Bacillus subtilis genome-scale metabolic model improves phenotype predictions and enables in silico design of polyγ-glutamic acid production strains. *Microbial Cell Factories* 18, 3, doi:10.1186/s12934-018-1052-2 (2019).

[110] Patel, V., Wu, Q., Chandrangsu, P. & Helmann, J. D. A metabolic checkpoint protein GlmR is important for diverting carbon into peptidoglycan biosynthesis in Bacillus subtilis. *PLOS Genetics* 14, e1007689, doi:10.1371/journal.pgen.1007689 (2018).

[111] Sonenshein, A. L. Control of key metabolic intersections in Bacillus subtilis. *Nature Reviews Microbiology* 5, 917-927, doi:10.1038/nrmicro1772 (2007).

[112] Brinsmade, S. R. et al. Hierarchical expression of genes controlled by the Bacillus subtilis global regulatory protein CodY. *Proceedings of the National Academy of Sciences* 111, 8227, doi:10.1073/pnas.1321308111 (2014).

AUTHOR CONTACT INFORMATION

Wenfa Ng
Master of Engineering
National University of Singapore, Singapore
Email: ngwenfa771@hotmail.com

INDEX

C

E

G

H

M

N

P

R

S